Topics in Electron Diffraction and
Microscopy of Materials

MICROSCOPY IN MATERIALS SCIENCE SERIES

Series Editors

B Cantor and M J Goringe

Other titles in the series

Electron Microscopy of Interfaces in Metals and Alloys
C T Forwood and L M Clarebrough

The Measurement of Grain Boundary Geometry
V Randle

Forthcoming topics to be covered by the series include

Quasicrystals
Catalysis
Magnetic Materials
Semiconductor Materials
Irradiated Materials
Ceramics
CBED
Orientation Imaging
STEM
Applications of HREM to Materials

*Please contact the Publisher for Materials Science,
Institute of Physics Publishing, Dirac House, Temple Back, Bristol BS1 6BE, UK
for more information*

MICROSCOPY IN MATERIALS SCIENCE SERIES

Topics in Electron Diffraction and Microscopy of Materials

P B Hirsch

Kt, MA, PhD, FRS

University of Oxford, UK

Institute of Physics Publishing
Bristol and Philadelphia

British Library Cataloguing-in-Publication Data

A catalogue record for this book is available from the British Library.

ISBN 0 7503 0538 X

Library of Congress Cataloging-in-Publication Data are available

Published by Institute of Physics Publishing, wholly owned by The Institute of Physics, London

Institute of Physics Publishing, Dirac House, Temple Back, Bristol BS1 6BE, UK

US Office: Institute of Physics Publishing, The Public Ledger Building, Suite 1035, 150 South Independence Mall West, Philadelphia, PA 19106, USA

Typeset in TeX by Digital by Design, Cheltenham
Printed in the UK by J W Arrowsmith Ltd, Bristol

CONTENTS

FOREWORD

Materials science has evolved as a crucial engineering discipline during the last 40 years. The basic approach of the materials scientist is to investigate the microstructure of a material, so as to optimize its manufacturing process and resulting properties for subsequent engineering use. Electron microscopy has proved to be by far the most powerful technique for examining and understanding material microstructure, and electron microscope methods are essential for developing new engineering materials of all types. The objective of this series of books is to provide overviews of the impact of electron microscopy and other microscopy techniques in different branches of materials science. The series is designed to be broadly based across the materials spectrum, including metals, ceramics, polymers and semiconductors, and to deal with the full range of available electron microscope and related techniques, such as electron diffraction, lattice imaging, scanning electron microscopy and microprobe analysis. The series is intended to be of interest to a wide variety of academic and industrial research scientists and engineers.

The present book began with a meeting held to mark the retirement of Professor M J Whelan. Nine eminent colleagues, under the editorship of Professor Sir Peter Hirsch, have produced a text on key aspects of the development and application of transmission electron microscopy and analysis to which Professor Whelan himself has made outstanding contributions during his long and distinguished career. M J Goringe was a PhD student under Whelan's supervision in Cambridge, and we were both colleagues of Whelan and Hirsch for many years in Oxford. It is a great privilege for us to be associated in this book with such eminent scientists and top-rank contributors of the individual chapters. Each chapter reviews and analyses a particular electron microscope technique and examines its application to problems in materials science.

B Cantor and M J Goringe
Series Editors

FRONTISPIECE

Whelan, Professor Michael John, MA, PhD, DPhil; FRS 1976; FInstP; Professor of Microscopy of Materials, Department of Materials, University of Oxford, 1992–97; Fellow of Linacre College, Oxford, since 1967; *b* 2 Nov. 1931; *s* of William Whelan and Ellen Pound. *Educ*: Farnborough Grammar Sch.; Gonville and Caius Coll., Cambridge. FInstP 1976. Fellow of Gonville and Caius Coll., 1958–66; Demonstrator in Physics, Univ. of Cambridge, 1961–65; Asst Dir. of Research in Physics, Univ. of Cambridge, 1965–66; Reader, Dept of Materials, Univ. of Oxford, 1966–92. C V Boys Prize, Institute of Physics, 1965; Victor G Macres Memorial Award, Microbeam Analysis Society, 1976; Hughes Medal of Royal Society, 1988; Distinguished Scientist Award of EMSA, 1998. Hon. Professor, Univ. of Sci. and Technology, Beijing, People's Republic of China, 1995–. *Publications*: (co-author) *Electron Microscopy of Thin Crystals*, 1965; *Worked Examples in Dislocations*, 1990; numerous papers in learned journals. *Recreation*: gardening. *Address*: Department of Materials, Parks Road, Oxford OX1 3PH, UK. *Tel*: Oxford (01865) 273700. *Home address*: 18 Salford Road, Old Marston, Oxford OX3 ORX, UK. *Tel*: Oxford (01865) 244556.

PREFACE

Over the last fifty years there have been tremendous advances in the development and application of electron optical techniques to the study of materials. These techniques are remarkable for their variety and great power. They can be used to probe the atomic and electronic structure of materials, some with atomic resolution, both inside the materials and at their surfaces. In order to interpret the images and diffraction patterns, the elastic and inelastic scattering processes of electrons in crystals must be understood.

This volume brings together contributions from authorities who have made seminal advances to the development of techniques and/or to the underlying theory of electron diffraction and inelastic electron scattering. The articles were presented at a symposium held in Oxford on 18 September 1997 in honour of Professor M J Whelan, FRS, on the occasion of his retirement from his post in the Department of Materials at Oxford University. Professor Whelan himself (or with collaborators) has made outstanding pioneering contributions to the development of the diffraction contrast technique for imaging defects in crystals, the theory of inelastic scattering and the technique of EELS, and, more recently, to the theory of backscattering of high energy electrons and of channelling patterns and to the theory of reflection high energy electron diffraction (RHEED) from surfaces. Most of the contributions to this volume cover aspects of current interest in all these areas, having as a common thread the dynamical theory of electron diffraction and inelastic scattering mechanisms. In addition there are articles on atomic resolution imaging using high energy and low energy electrons, exploiting the different physical mechanisms which control the two cases. There is also a historical account of the early days of the development of diffraction contrast imaging of defects in crystals, in which Professor Whelan played such a leading role. All the contributors have either collaborated with Professor Whelan or have been close scientific colleagues with common interests.

So this monograph, consisting of a set of authoritative articles on various aspects of electron microscopy and diffraction of interest to the materials scientist, is also to be seen as a tribute to Professor Whelan. His influence and impact in the field have been immense, and the contributors wish to express their appreciation on behalf of the many scientists all over the world who have enjoyed his friendship and his stimulation of their scientific work.

P B Hirsch
Editor

AUTHORS' ADDRESSES

G A Botton Materials Technology Laboratory, CANMET, NRCan, 568 Booth Street, Ottawa, Ontario K1A 0G1, Canada

D J H Cockayne Australian Key Centre for Microscopy and Microanalysis, University of Sydney, NSW, Australia

S L Dudarev Department of Materials, University of Oxford, Parks Road, Oxford OX1 3PH, UK

H Hashimoto Okayama University of Science, 1-1, Ridaicho, Okayama 700-0005, Japan

P B Hirsch Department of Materials, University of Oxford, Parks Road, Oxford OX1 3PH, UK

A Howie Cavendish Laboratory, University of Cambridge, Madingley Road, Cambridge CB3 0HE, UK

X Huang Department of Physics, Arizona State University, Tempe, AZ 85287, USA

C J Humphreys Department of Materials Science and Metallurgy, University of Cambridge, Pembroke Street, Cambridge CB2 3QZ, UK

A F Moodie Department of Applied Physics, RMIT, GPO Box 2476V, Melbourne 3001, Australia

Lian-Mao Peng Beijing Laboratory of Electron Microscopy, Institute of Physics and Center for Condensed Matter Physics, Chinese Academy of Sciences, PO Box 2724, Beijing 100080, People's Republic of China

J C H Spence Department of Physics, Arizona State University, Tempe, AZ 85287, USA

D Taylor Institute of Molecular Biophysics, Florida State University, Tallahassee, FL 32306-4380, USA

K Taylor Institute of Molecular Biophysics, Florida State University, Tallahassee, FL 32306-4380, USA

U Weierstall Department of Physics, Arizona State University, Tempe, AZ 85287, USA

1

EARLY DAYS OF DIFFRACTION CONTRAST TRANSMISSION ELECTRON MICROSCOPY

P B Hirsch

1.1 INTRODUCTION

In the 1940s and 1950s solid state physicists made major advances in the development of the theory of dislocations in crystals, and in its application to crystal growth phenomena and mechanical properties (see e.g. [1]). This stimulated the search for techniques to observe dislocations directly under the microscope. Following Frank's suggestion that crystals can grow under low supersaturation conditions if they contain screw dislocations intersecting the surface [2], Griffin reported in 1950 the existence of the predicted growth spirals on the surface of beryl crystals [3]. Similar observations were made subsequently in many other types of crystal (see e.g. [4]). Then etch pit techniques were developed, which revealed dislocations at the surface in a variety of materials including metals and alloys. These could be used to check certain predictions of dislocation theory such as Frank's formula for the dependence of misorientation at a low angle boundary on the spacing of the edge dislocations. Vogel [5] reported in 1955 good quantitative agreement for low angle tilt boundaries in single crystals of Ge, using x-ray diffraction to determine the angular rotation. Such experiments also gave support to the assumption that the etch pits did indeed correspond to individual dislocations. Another method used was that of preferential precipitation at dislocations. Hedges and Mitchell [6] used this technique to reveal dislocation networks inside crystals of AgBr, the photolytic silver after annealing and exposure to light precipitating on the dislocation lines. For metals and alloys however dislocations could only be seen at the surface (see e.g. [7]). The evidence for dislocations and the various techniques used in this period were reviewed in 1956 [8].

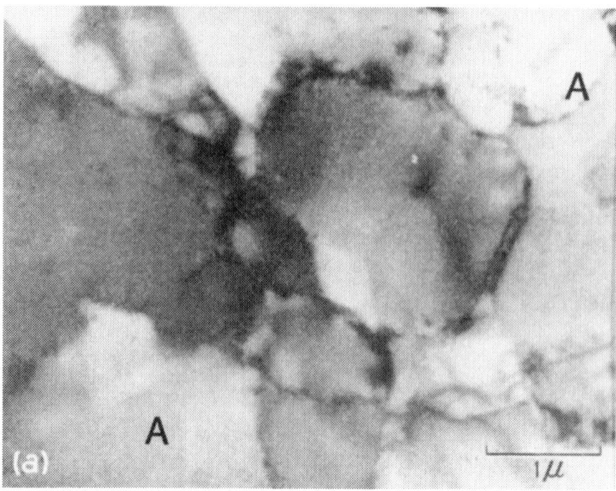

Figure 1.1 TEM image of subgrains in beaten aluminium foil, thinned by electropolishing after deformation. The bands seen at A may be slip traces (from [13], [48], courtesy Elsevier Science).

In a parallel development, W L Bragg in Cambridge suggested an experiment to settle a controversy about whether x-ray line broadening from work-hardened metals is due to small particle size effects or internal strains. The idea was to obtain an x-ray diffraction pattern from such a small volume of metal that if the structure consisted of small subgrains a spotty rather than continuous diffraction ring would be obtained. Using an intense x-ray beam from a specially constructed fine focus rotating anode tube, collimated down to $\leqslant 10$ μm diameter using glass capillaries, spotty rings were indeed obtained from cold-worked Al, and the subgrain size was deduced to be 2 μm [9]. This microbeam technique was also used successfully to determine the microstructure of other deformed metals, including iron [10], and generally these experiments showed that as the subgrain size was rather large in these cases, the line broadening was mainly due to elastic distortion of the subgrains. For some metals, however, in particular beaten gold foil, it proved impossible to obtain diffraction rings with clearly resolved spots [11].

About the same time, however, Heidenreich [12, 13] had published transmission electron microscope images of electrolytically thinned specimens of annealed and beaten foils of Al. He interpreted extinction contours in the annealed samples in terms of the dynamical theory of electron diffraction, and he found that the structure of beaten Al foil consisted of subgrains about 2 μm in diameter, in excellent agreement with the x-ray microbeam results. Figure 1.1 is one of Heidenreich's micrographs taken from [13]. Although Heidenreich did not report any observations of individual dislocations, his seminal work can

Figure 1.2 TEM image of beaten gold foil, taken by J W Menter in an AEI EM3 electron microscope at 70 kV. The bands of contrast were thought to arise from the phase change in the diffracted beam across a stacking fault (from [48], courtesy Elsevier Science).

be regarded as laying the foundation for the diffraction contrast technique, and indicated the way forward.

Initially, however, the Cambridge group used the electron diffraction facility available in electron microscopes to provide very fine and intense electron beams to illuminate specimens with smaller particle sizes. Spotty rings were in fact obtained in transmission through beaten gold foil [14], but Menter, who operated the microscope, also produced micrographs, one of which is shown in figure 1.2. The foils have a pronounced cube texture and the white and dark bands which can be seen on the micrograph are parallel to the traces of {111} planes. This coupled with the fact that the diffraction patterns showed strong streaking along ⟨111⟩ directions, suggested that these narrow bands of contrast corresponded to stacking faults, which gave contrast arising from the phase change suffered by the electron beam from the displacement of the crystal across the fault. This led to the idea that dislocations in f.c.c. lattices which were thought to be dissociated into partial dislocations bounding a stacking fault, as proposed by Heidenreich and Shockley in 1948 [15], could be imaged

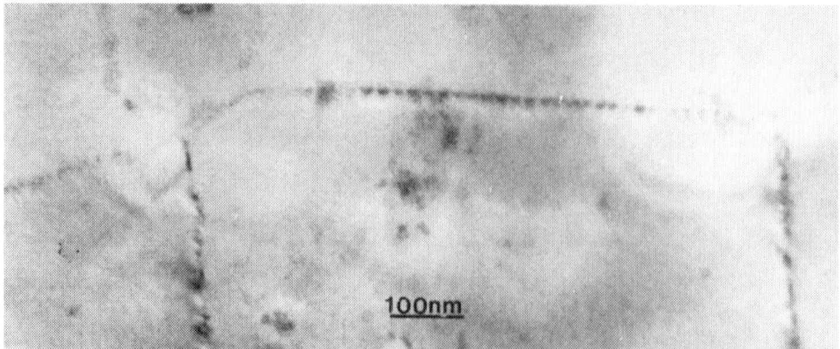

Figure 1.3 TEM image of a thin foil of aluminium beaten and then annealed and thinned by etching showing a subgrain boundary consisting of dislocations spaced uniformly at a spacing of 17.5 nm (from [18], courtesy Taylor and Francis).

through the effect of the stacking-fault ribbon. Furthermore, the widths of the bands of contrast on the images in beaten gold foil (figure 1.2) suggested that it might be possible to 'see through' metal foils 1000 Å thick or more. Electron microscopists at the time were rather sceptical about this, on the evidence from transmission through fine grained polycrystalline foils, but there was the hope that an anomalous transmission effect, similar to the Borrmann effect in x-ray diffraction [16], might also operate in electron diffraction. Armed with these ideas the decision was made to explore the possibility of observing dislocations directly in thin metal foils by diffraction contrast in the electron microscope. All that was needed was a good research student and a good microscope. As fate would have it both materialized at about the same time.

1.2   FIRST EXPERIMENTS

M J Whelan joined my group in the Cavendish Laboratory in Cambridge as a research student in October 1954 to work on this project. In the same year Cosslett in the Electron Microscopy Group in the Cavendish Laboratory took delivery of one of the new Siemens Elmiskop I microscopes. This instrument was operated by R W (Bob) Horne and used primarily for biological applications, but time was made available to the dislocation project. Much of the first year was spent building an ion-beam bombardment apparatus for thinning the foils following previous work by Castaing [17]. However, application to gold foils produced unsatisfactory specimens with large etched holes. In any case it seemed likely that the chances of success would be better with partially annealed structures such as those in Al studied by Heidenreich, than with the very complex structures in Au. Whelan managed to produce suitable specimens of Al about

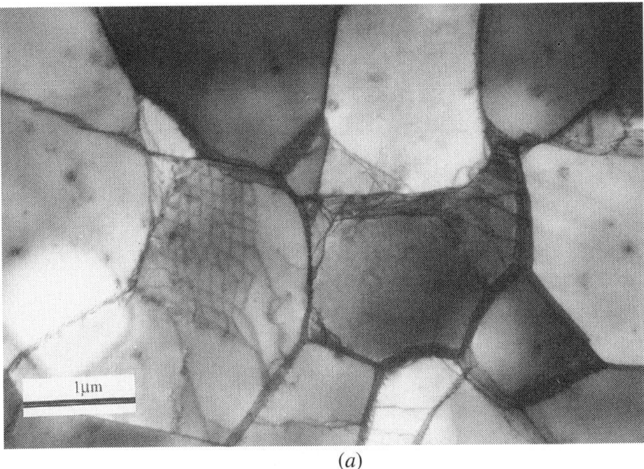

(*a*)

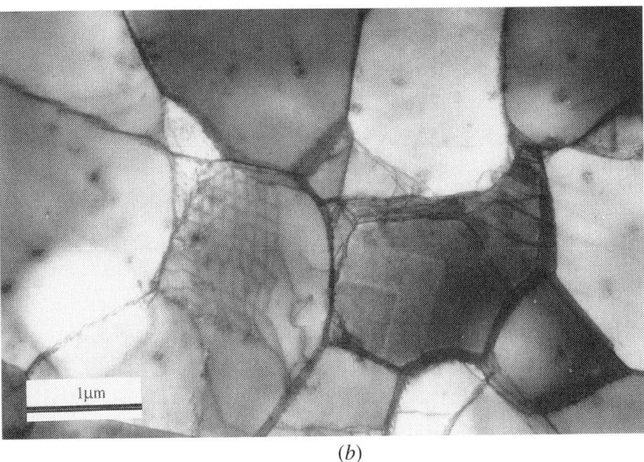

(*b*)

Figure 1.4 TEM images of the same area in a thin foil of beaten and annealed aluminium, viewed in successive exposures, taken by M J Whelan in 1956 showing dislocations in subgrain boundaries, including a network nearly parallel to the foil plane. Between the two exposures a dislocation has slipped across a subgrain leaving behind a cross-slip trace in the lower image (see [18], [29] and [48], courtesy Institute of Materials).

1000 Å thick from beaten and subsequently annealed Al foil by etching in dilute HF solution. Images such as that of figure 1.3, looking like subgrain boundaries consisting of individual dislocations, were obtained, but there was some doubt whether the array of dots could not be due to a moiré effect.

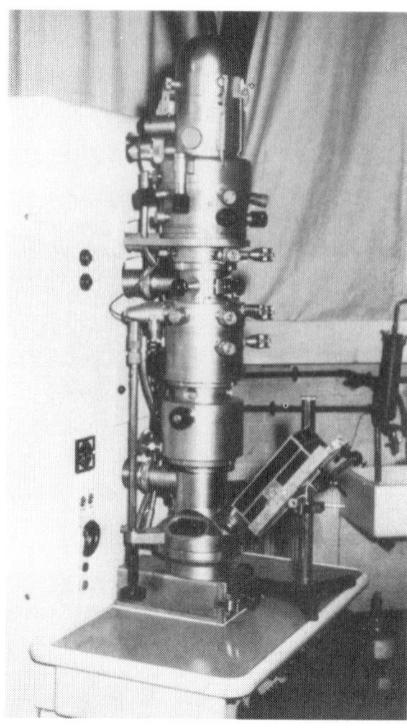

Figure 1.5 Siemens Elmiskop I installed at the Cavendish Laboratory, Cambridge in 1954, with the NPL ciné camera used to film moving dislocations (from [49], courtesy John Wiley and Sons).

These observations were made in a mode of operation of the Elmiskop I in which switching from microscopy to diffraction was easy. However, under these conditions the double condenser system could not be used. On 3 May, 1956 Bob Horne used the microscope in the 'high resolution mode' with the double condenser system, and the condenser aperture was removed to increase the beam intensity. The previously stationary lines began to bow out and move, leaving behind traces parallel to (111) slip planes, which, although they gradually faded, persisted long enough to be recorded. All at once the doubts were dispelled—we were observing dislocations and their movement. Figure 1.4 consists of two consecutive images taken by Whelan from the same area of foil, showing subgrains, a dislocation network, and a slip trace in the second image which shows that a dislocation had moved and cross-slipped between the two exposures. The results were published in the July issue of the *Philosophical Magazine* [18]. A movie film was made of the movement of dislocations in Al by photographing the tilted image on the fluorescent screen of the microscope with a ciné camera (borrowed from the National Physical Laboratory), through the central of the three windows on the Elmiskop I; the arrangement is shown in figure 1.5.

Although the original ideas which led to these experiments were later shown to be valid, the contrast of the dislocation images in Al is of course due to the bending of the lattice planes, and not due to the stacking-fault ribbon, since the width of the latter is negligibly small for this metal. Similarly, the possibility of anomalous transmission was largely irrelevant in these first experiments on thin foils of Al.

Independently of these experiments Walter Bollmann at the Battelle Institute in Geneva was working on a project to try to image dislocations directly in thin foils of stainless steel in the electron microscope. This work was suggested by Dr Siegfried (who was then the Head of the Metallurgy Section), who had been stimulated by Castaing's TEM studies of Al–4% Cu alloys [17]. Bollmann developed a very successful electropolishing technique for preparing thin foils of stainless steel, and observed dislocation lines crossing the foil [19]. However, his Philips EM100 electron microscope lacked the advantage of a double condenser system, and the dislocations did not move. Some of his images showed ribbons with fringes which we interpreted tentatively as arising from stacking faults. It was clear that stainless steel would be a very useful alloy for detailed studies of stacking faults and dislocations, and a collaboration was agreed to carry out experiments on Bollmann's stainless steel specimens on the Siemens Elmiskop 1.

1.3   THEORY OF IMAGE CONTRAST FROM STACKING FAULTS AND EXPERIMENTS ON STAINLESS STEEL

The availability of suitable specimens gave the impetus to the development of the theory of image contrast from stacking faults. Whelan worked out the two beam dynamical theory of image contrast from inclined stacking faults in the Christmas vacation in 1956, and we both worked out independently the kinematical theory of contrast, and introduced the 'column approximation'. The contrast arises from the displacement of R across the fault, which results in a phase angle $\alpha = 2\pi g \cdot R$, where g is the reciprocal lattice vector for the operating reflection. Detailed experiments were carried out by Whelan on stainless steel and the agreement between theory and experiment was generally good, but there were some discrepancies which were resolved when absorption was taken into account later (see section 1.4). These experiments were quite difficult because no goniometer stage was available and tilting was limited to one axis using the stereo holder. Nevertheless this work provided detailed understanding of the origin and nature of the contrast from stacking faults [20, 21], and provided confirmation of the original idea. This was the first theoretical treatment of the image contrast from any crystal defect and proved seminal. In his PhD thesis under suggestions for further work, Whelan wrote

1. Extension of the theory to crystals containing imperfections other than stacking faults. It is possible to develop, at least formally, a generalized

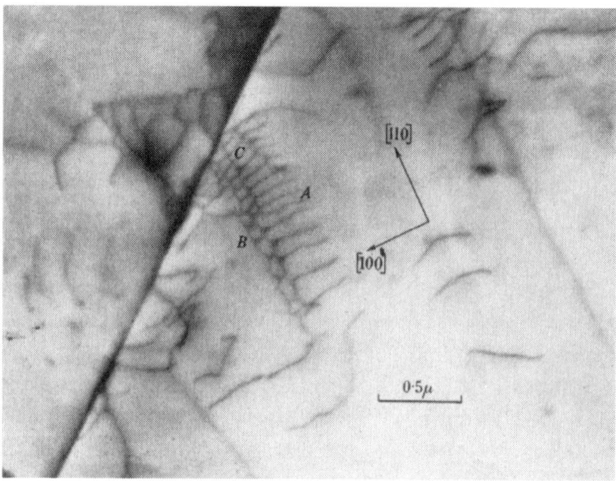

Figure 1.6 Dislocation network on a (111) plane in slightly deformed stainless steel, showing extended and contracted nodes (from [22], courtesy Royal Society).

dynamical theory by considering the phase factor α to be a continuous variable. The resulting equations may possibly be solved numerically. In this way it should be possible to give a quantitative theory of other contrast effects such as at dislocations, slip traces, and grain boundaries.

2. Extension of the theory to include absorption. Formally this can be included in the theory by introducing complex lattice potentials. However, at present the nature of these is unknown.

The stainless steel specimens were useful not only for experiments carried out to test the contrast theory, but also for studying the behaviour and interaction of dissociated dislocations [22]. Observations were made of dislocations piled up against grain boundaries, and of interactions between dislocations on different slip planes and with different Burgers vectors. Figure 1.6 taken from [22] shows extended and contracted nodes in dislocation networks of dissociated dislocations, as predicted by dislocation theory. Using the double condenser system, the dislocations could be made to move; figure 1.7 shows successive images of the same area showing a separation of the partial dislocations for a dislocation at A, followed by movement of the front partial to the boundary. The figure also shows the strong pinning of the dislocation by the surface oxide film, and dislocation bowing between the pinning points. Many other interesting observations were made, and a ciné film of the motion of the dislocations in stainless steel was produced. This film and that on Al were shown in many laboratories all over the world.

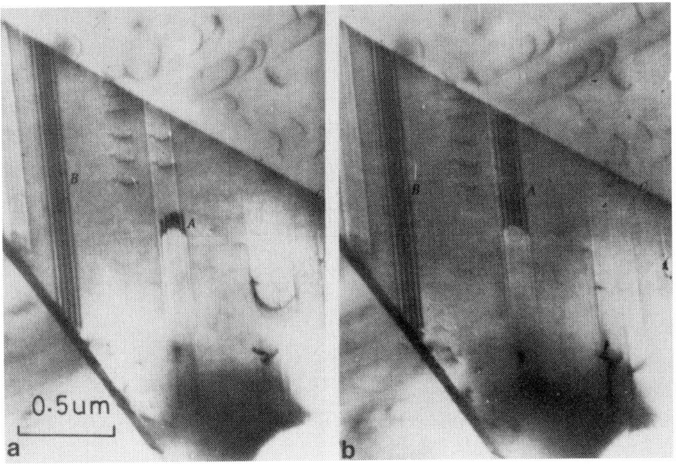

Figure 1.7 Dissociation of dislocations in stainless steel. Images (*a*), (*b*) are taken in successive exposures. The front partial dislocation at A in (*a*) moves away in (*b*) to a twin boundary at C leaving behind a wide stacking fault showing characteristic fringes. Note the bowing of the dislocations due to pinning at the surface (from [22], courtesy Royal Society).

A further detailed study was carried out by Whelan on the interactions between intersecting dislocations, and several predictions of dislocation theory about the nature of the reaction products were confirmed, including Lomer–Cottrell locks and stair-rod dislocations [23]. This work also provided the first method of determining stacking-fault energy directly from dislocation configurations, namely from curvature of the extended nodes. A very important conclusion arising out of this study was that the three-dimensional nature of dislocation interactions had to be taken into account, and that, unlike in the two-dimensional theoretical models, dislocation locks had finite lengths. This in turn led to the idea that the important interactions in work-hardening were attractive interactions between intersecting dislocations, thus providing a new slant to the 'forest' theory of flow stress.

1.4 CONTRAST THEORY FROM DISLOCATIONS AND ANOMALOUS ABSORPTION

The next development of image contrast theory was the kinematical theory of diffraction contrast from dislocations, which was based on the column approximation and made use of amplitude phase diagrams. This treatment established the 'invisibility criterion' used for Burgers vector analysis. The work was published in 1960 [24], jointly with Howie who had joined the

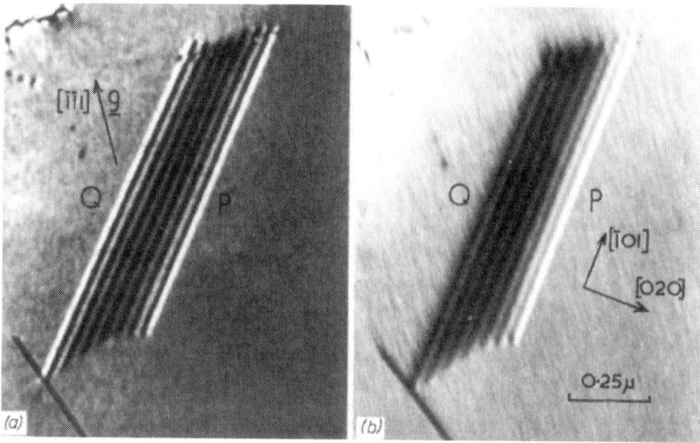

Figure 1.8   Bright-field (a) and dark-field (b) images of a stacking fault in Cu–7% Al alloy.   The bright-field image fringes are symmetrical about the foil centre, but have reduced visibility near the centre.   The dark-field image is asymmetrical about the centre (from [28], courtesy Royal Society).

Metal Physics Group in Cambridge in 1957.   Howie and Whelan then went on to develop the dynamical theory of image contrast from dislocations, and formulated the famous Howie–Whelan equations [25, 26].

Another important development was the introduction of absorption into the dynamical theory through an imaginary part of the scattering potential.   The theory for stacking-fault images was developed and compared with experiment during a stimulating visit by Professor H Hashimoto in 1959–1960 [27, 28]. This theory explained the fading of the thickness fringes from inclined faults near the middle of the foils, which had been noticed previously, and also the facts that bright- and dark-field images are symmetrical and asymmetrical, respectively, about the centre.   This is shown in figure 1.8 taken from [28]. This treatment also explained the important observation that for certain crystal orientations in thick crystal regions the electrons suffer anomalous absorption, while for others there is 'anomalous transmission'.   Thus, as originally thought, there is an effect for electron propagation similar to the 'Borrmann' effect for x-rays [16], although the main absorption mechanisms are rather different in the two cases—being thermal diffuse scattering for electrons and the ejection of electrons from the inner shells of the atoms for x-rays.   While most computations were carried out using the two-beam approximation, the Howie–Whelan theory is a general many-beam scattering matrix formulation, taking into account absorption, which can be generally applied to defects in crystals. By this time ($\sim$1962) the theory of diffraction contrast images from defects, based on the column approximation, had been developed to a stage at which

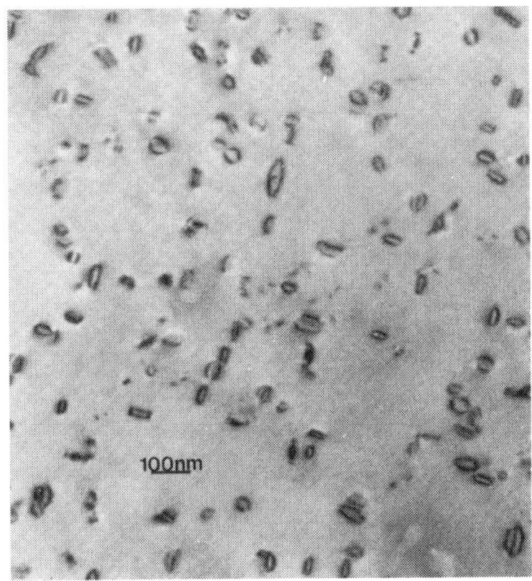

Figure 1.9 Dislocation loops in pure aluminium quenched from 560 °C and aged at room temperature (see [30]).

it could be applied with confidence to the characterization of many types of defect.

1.5 IMPACT OF 'DIFFRACTION CONTRAST' TECHNIQUE IN METALLURGY, AND EARLY APPLICATIONS

The advent of the 'diffraction contrast' technique did much to convince the more sceptical metallurgists at the time that dislocations in metals were real and important, and not just the figment of the fertile imagination of solid state physicists. It was a case of 'seeing is believing'. In the words of Sir Alan Cottrell [29], 'From that day onwards, dislocations ceased to be a theory and became observed features of crystalline materials taking their place alongside grain boundaries, twins, slip bands and alloy particles, as things to be seen and studied under the microscope. A new door had opened in materials science'.

By the mid-1950s dislocation theory had been developed to a remarkable degree of detail, and there was much scope for experimental tests of the models. Whelan's work on dislocation interactions in stainless steel [23] is a good example of this, although, as mentioned above, it also highlighted the essential three-dimensional nature of the dislocation arrangement. For quenched metals the technique provided decisive evidence for the existence of prismatic loops [30], which was one of several models suggested for the annealing of vacancies. A micrograph taken from the work on quenched Al [30] is shown in figure 1.9.

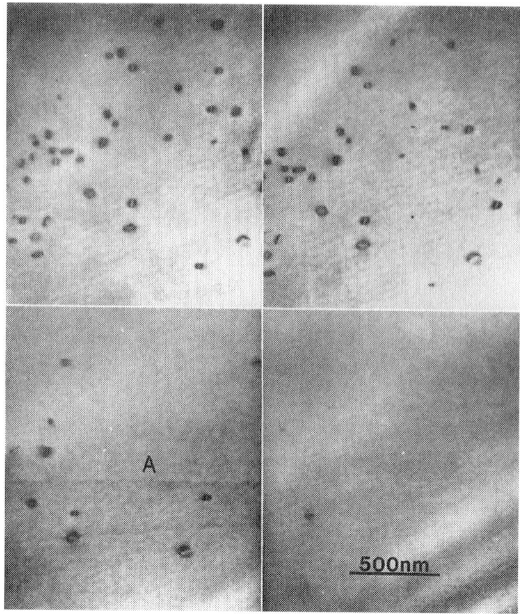

Figure 1.10   Sequence of micrographs of the same area showing the annealing of loops in quenched aluminium examined at $\sim$190 °C on a heating stage in the electron microscope. The loops can be seen to shrink by climb and eventually to disappear (from [33], courtesy Taylor and Francis).

Tetrahedra of stacking faults first observed in 1959 in quenched Au [31] are an example of a three-dimensional defect not previously considered. Electron microscope observations of irradiated metals provided evidence for the first time for a mechanism of recovery of displacement zones into dislocation loops [32]. These early experiments opened up a whole new field and there is little doubt that over the years the identification by TEM of the nature of defects in irradiated materials played a crucial role in the development of understanding of the processes of irradiation damage in metals and alloys.

The possibility of studying dynamic processes directly under controlled conditions *in situ* in the microscope led to the development of heating, cooling and straining stages, and Whelan himself designed a heating stage in 1957 which was used in a pioneering study of the annealing of dislocation loops in quenched Al, in which the process of climb and shrinkage of loops by diffusion was observed directly [33]. Figure 1.10 is a set of micrographs illustrating this process, taken from [33].

1.6   SUBSEQUENT DEVELOPMENTS

After the initial phase from 1956 to the early 1960s, there followed a period of development in various directions and rapidly increasing application of the technique in laboratories all over the world. The 'weak beam technique' [34] developed in 1969 represented a step function improvement (by a factor of ten) of the resolution at which defects could be imaged by diffraction contrast (down to about 1.5 nm), and this has proved to be an exceedingly powerful method for studying the detailed geometry of quite complex extended defects (for a review see chapter 2). In the 1970s high voltage electron microscopes (say 1 MeV) became commercially available, and this led to further development of special stages for *in situ* observations under controlled conditions, and this resulted in an upsurge in this kind of work in thicker specimens (for a review see [35]). As an example, *in situ* straining experiments on b.c.c. metals showed quite clearly and directly that the plasticity of these metals at low temperatures is controlled by the resistance of the lattice to the motion of screw dislocations (the Peierls force (for a review see [36])).

Much progress has been made over the years in the theory of inelastic scattering, and the effect of inelastic scattering on images and of course on absorption. Some of Whelan's contributions in this field are published in [37] and [38]. Contributions to the problem of preservation of image contrast after various types of inelastic scattering are described in [39] and [40]. More recently treatments involving both diffraction and multiple inelastic scattering have been developed, which can be applied to backscattering of high energy electrons, channelling patterns and electron channelling contrast imaging [41, 42]; these topics are discussed in chapter 8.

One of the most important developments has been the field of electron energy loss spectroscopy (EELS). Cundy, Metherell and Whelan [43] built an energy analysing microscope which was used in pioneering studies of segregation and precipitation in Al alloys [44], using plasmon losses. Subsequently, the emphasis shifted to the higher energy losses associated with inner shell excitations, and the relation between energy loss spectra and band structure [45, 46]. These were important steps in the development of EELS which today is a powerful microanalytical technique with high spatial resolution and high sensitivity to the electronic structure. This topic is discussed in chapters 5 and 6, and the EELS technique is reviewed in the book by Egerton [47].

1.7   CONCLUSIONS

In this chapter I have concentrated on the early development of the 'diffraction contrast' technique and referred only briefly to those subsequent developments to which Professor Whelan has made many contributions. Further details about the early developments are given in [48] and [49].

There have of course been many other very important advances in transmission electron microscopy, notably in respect of instrumental resolution, high resolution electron microscopy and atomic imaging (see chapter 4), and other techniques such as convergent beam electron diffraction (CBED) (see chapters 3 and 5). Today electron microscopy, both transmission and scanning, and the associated analytical techniques provide the materials scientist with an armoury of unprecedented power and versatility for the development of materials, and the 'diffraction contrast' technique, to the development of which Professor Whelan has made so many pioneering contributions, forms an important part of this armoury.

REFERENCES

[1]   Cottrell A H 1953 *Dislocations and Plastic Flow in Crystals* (Oxford: Clarendon)
[2]   Burton W K, Cabrera N and Frank F C 1949 *Nature* **163** 398
[3]   Griffin L J 1950 *Phil. Mag.* **41** 196
[4]   Verma A R 1953 *Crystal Growth and Dislocations* (London: Butterworth)
[5]   Vogel F L 1955 *Acta Metall.* **3** 245
[6]   Hedges J M and Mitchell J W 1953 *Phil. Mag.* **44** 223; 1953 *Phil. Mag.* **44** 357
[7]   Wilsdorf H and Kuhlmann-Wilsdorf D 1954 *Phil. Mag.* **45** 1096
[8]   Hirsch P B 1956 *Mosaic Structure (Progress in Metal Physics VI)* (London: Pergamon) p 236
[9]   Hirsch P B, Kellar J N and Thorp J S 1950 *Nature* **165** 554
[10]   Gay P and Kelly A 1953 *Acta Crystallogr.* **6** 165; 1953 *Acta Crystallogr.* **6** 172
[11]   Kelly A 1953 *PhD Thesis* University of Cambridge
[12]   Heidenreich R D 1949 *J. Appl. Phys.* **20** 993
[13]   Heidenreich R D 1951 *Bell Syst. Tech. J.* **30** 867
[14]   Hirsch P B, Kelly A and Menter J W 1955 *Proc. Phys. Soc.* B **68** 1132
[15]   Heidenreich R D and Shockley W 1948 *Report Conf. on the Strength of Solids (Bristol, 1947)* (London: Physical Society) p 57
[16]   Borrmann G 1950 *Physica* **127** 297
[17]   Castaing R 1955 *Rev. Métall.* **52** 669
[18]   Hirsch P B, Horne R W and Whelan M J 1956 *Phil. Mag.* **1** 677
[19]   Bollmann W 1956 *Phys. Rev.* **103** 1588
[20]   Whelan M J and Hirsch P B 1957 *Phil. Mag.* **2** 1121
[21]   Whelan M J and Hirsch P B 1957 *Phil. Mag.* **2** 1303
[22]   Whelan M J, Hirsch P B, Horne R W and Bollmann W 1957 *Proc. R. Soc.* A **240** 524
[23]   Whelan M J 1958 *Proc. R. Soc.* A **249** 114

[24] Hirsch P B, Howie A and Whelan M J 1960 *Phil. Trans. R. Soc.* A **252** 499
[25] Howie A and Whelan M J 1961 *Proc. R. Soc.* A **263** 217
[26] Howie A and Whelan M J 1962 *Proc. R. Soc.* A **267** 206
[27] Hashimoto H, Howie A and Whelan M J 1960 *Phil. Mag.* **5** 967
[28] Hashimoto H, Howie A and Whelan M J 1962 *Proc. R. Soc.* A **269** 80
[29] Cottrell A H 1997 *Interdisciplinary Sci. Rev.* **22** 318
[30] Hirsch P B, Silcox J, Smallman R E and Westmacott K H 1958 *Phil. Mag.* **3** 897
[31] Silcox J and Hirsch P B 1959 *Phil. Mag.* **4** 72
[32] Silcox J and Hirsch P B 1959 *Phil. Mag.* **4** 1356
[33] Silcox J and Whelan M J 1960 *Phil. Mag.* **5** 1
[34] Cockayne D J H, Ray I L F and Whelan M J 1969 *Phil. Mag.* **20** 1265
[35] Imura T, Maruse S and Suzuki T 1986 *Japan. J. Electron Microsc.* **35** (Supplement)
[36] Louchet F, Kubin L P and Vesely D 1979 *Phil. Mag.* A **39** 433
[37] Whelan M J 1965 *J. Appl. Phys.* **36** 2099
[38] Whelan M J 1965 *J. Appl. Phys.* **36** 2103
[39] Humphreys C J and Whelan 1969 *Phil. Mag.* **20** 165
[40] Rossouw C J and Whelan M J 1981 *Ultramicroscopy* **6** 53
[41] Dudarev S L, Peng L M and Whelan M J 1993 *Phys. Rev.* B **48** 13 408
[42] Dudarev S L, Rez P and Whelan M J 1995 *Phys. Rev.* B **51** 3397
[43] Cundy S L, Metherell A J F and Whelan M J 1966 *J. Sci. Instrum.* **43** 712
[44] Cundy S L, Metherell A J F, Whelan M J, Unwin P N T and Nicholson R B 1968 *Proc. R. Soc.* A **307** 267
[45] Egerton R F and Whelan M J 1974 *Phil. Mag.* **30** 739
[46] Egerton R F and Whelan M J 1974 *Electron Spectrosc.* **3** 232
[47] Egerton R F 1996 *Electron-Energy-Loss Spectroscopy in the Electron Microscope* (New York: Plenum)
[48] Hirsch P B 1986 *Mater. Sci. Eng.* **84** 1
[49] Whelan M J 1986 *J. Electron Microsc. Tech.* **3** 109

2

APPLICATIONS OF THE WEAK BEAM TECHNIQUE OF ELECTRON MICROSCOPY

David J H Cockayne

2.1   INTRODUCTION

The weak beam technique of electron microscopy [1] is now firmly established as a tool for investigating the geometry and nature of a wide range of crystalline defects. A review of the literature shows over 1000 publications in which the technique has been used; consequently any review of the applications of the technique must be severely selective, and this review is no exception.

The origin of the technique was the need to have a means for investigating the structure of dislocations at close to the resolution limit of the electron microscope. Although the standard high resolution technique (HREM) of interfering a number of diffracted beams gives images of dislocations at high resolution, generally the dislocation must be viewed along its line, and the length of line sampled is then so short ($\sim$10 nm) that the representative nature of the imaged dislocation is in question. Moreover, variations in the geometry along the line cannot be investigated by HREM.

The alternative approach of using either bright- or dark-field images of crystalline defects imaged with strongly excited Bragg reflections requires a great deal of image interpretation, largely because dynamical scattering and the extent of the strain field produce broad images with complex contrast variations. This interpretation can become second nature to the experienced microscopist (e.g. the recognition of intensity striations as coming from stacking faults; symmetry of the image to differentiate dipoles from single dislocations), but the complicated images can mask complex defect geometries. Even in the case of the simple dislocation line, the image is broad ($\simeq$15 nm wide), with intensity variations across the image due to changes in dislocation depth or foil thickness.

16

The pioneering work of Whelan (to whom this chapter is dedicated) and his colleagues tackled the problem of interpreting the image by providing the mathematical basis for calculating the intensity at each point in an image (for a comprehensive review see [2]). This allowed image profiles and image symmetries to be used to differentiate between different kinds of defect— dipoles and dislocations, intrinsic and extrinsic stacking faults, edge and screw dislocations, etc. Head [3] developed a very useful method of rapidly calculating and simulating the entire image for defects inclined in the foil, so that the variation of image intensity with dislocation depth could be displayed as a grey-scale image. This image could then be used for comparison with an experimental image, and so differentiate visually between possible alternative defect models.

In general, these techniques were used with strongly excited Bragg reflections, to maximize image contrast and defect visibility. The shortcoming of the images for investigating the detail of defect geometries was that the images were not sensitive, in any easily interpretable way, to details of defect geometries. For example dark-field images of dislocations taken with a strongly excited Bragg reflection have image widths of approximately 8–12 nm. Consequently two dislocations which are closer together than 10 nm (e.g. dislocations in f.c.c. metals dissociated into two Shockley partial dislocations) cannot easily be resolved. Moreover, the image peak position varies with dislocation depth and foil thickness to such an extent that it is very difficult to determine the position of the dislocation cores with accuracy. In the case of defects with more complex geometry (e.g. small loops, coherent precipitates), the relationship between the image and the defect geometry is extremely complicated. Image simulation is an aid, but even then possible alternative geometries must be guessed if simulations are to be performed, and the correct geometry might be overlooked.

The weak beam technique was developed to overcome many of these problems by producing images of dislocations which are narrow, and which accurately define the positions of dislocation cores. It was demonstrated that, to a good approximation, for a weak beam image taken with the reflection g, the image peak occurs for a column in the crystal within which the displacement field R bends the lattice planes into the Bragg orientation at the turning point of $g \cdot \mathrm{d}R/\mathrm{d}z$. This physical picture (figure 2.1) leads to a simple understanding of weak beam imaging.

This result is supported mathematically by the kinematic theory of dislocation contrast, where, for $g \cdot b = 1$ or 2, a dislocation shows one narrow image peak, close to the projection of the dislocation core (figure 2.2). For screw dislocations, the figure shows that for $g \cdot b = 2$ the image peak lies to one side of the core at a distance $\simeq 1/(2\pi s)$ where s is the distance of the Ewald sphere from the reciprocal lattice point. It is immediately evident from this diagram that, for $|s| \geqslant 2 \times 10^{-2} \ \text{Å}^{-1}$, this will give a narrow image peak ($\simeq 1.5$ nm wide) at a distance of approximately 1 nm from the dislocation core.

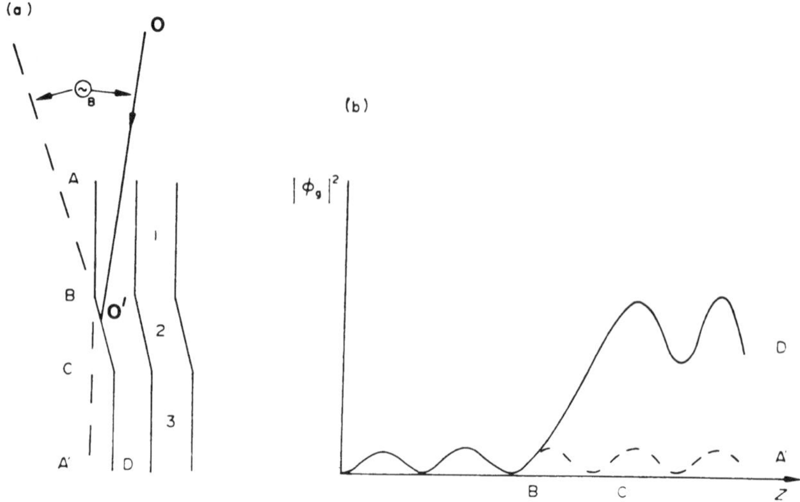

Figure 2.1   A diagrammatic representation of the origin of an image peak in a weak beam image. $|\phi_g|^2$ is the diffracted intensity. ϑ_B is the Bragg angle. OO′ is the direction of the incident beam.

In establishing the weak beam technique, the critical question was what, if any, *experimental* configuration would achieve the narrow image peak and a well defined relationship between the peak position and the position of the dislocation, as predicted by the kinematic theory. The factors which act against the kinematic prediction include the effect of dynamical scattering, the effect of multiple Bragg beams, the assumption of the column approximation and the influence of variations in dislocation depth and foil thickness on the image. The weak beam technique defined the conditions under which these effects could be minimized.

The early papers which developed the technique [1, 4], established experimental procedures which would allow the position of the dislocation cores, and their separations in the case of dissociated dislocations, to be determined to an accuracy better than 1 nm. These procedures are

1. $g \cdot b = 1$ or 2 (to ensure that each dislocation core gives rise to one, and only one, image peak),
2. $|s| \geqslant 2 \times 10^{-2}$ Å^{-1} (to ensure that the image peak width is as narrow as possible, consistent with being able to obtain an image given the weak image intensity),
3. no other reflection has $s \simeq 0$ (to ensure that scattering into the imaging reflection occurs only from the incident beam) and
4. $|w| = |s\xi| \geqslant 5$ (to obtain only one strongly excited Bloch wave [4]) (ξ is the extinction distance).

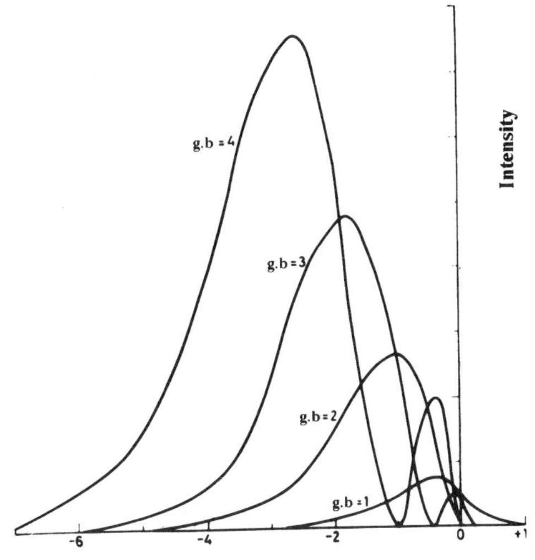

Figure 2.2 The kinematical image profile of a screw dislocation (after [62]).

If the above conditions are broken, the resulting images can still show fine image detail, which in turn can be used to investigate defect structure, but the ease of interpretation of the images might be lost.

A comprehensive review of the weak beam technique has been given elsewhere [5], and, in the following, a review is given of some of the areas of its application.

2.2   DISLOCATION DISSOCIATION

The earliest weak beam experiments were concerned with revealing the individual Shockley partial dislocations in dissociated dislocations. To achieve this, it was necessary to prove that weak beam images would produce only *one* image peak for *undissociated* dislocations. It is of historical interest that aluminium was considered an inappropriate material for this purpose, because theoretical studies suggested that dislocations in aluminium had a dissociation width of $\simeq 1.5$ nm; silicon was chosen because many believed that dislocations in that material were undissociated. When weak beam images of Si showed two image peaks, there was consternation, but a series of experiments [6] taken to reveal partial dislocations and stacking faults (e.g. figure 2.3) demonstrated that dislocations in silicon are dissociated, and this proved the technique.

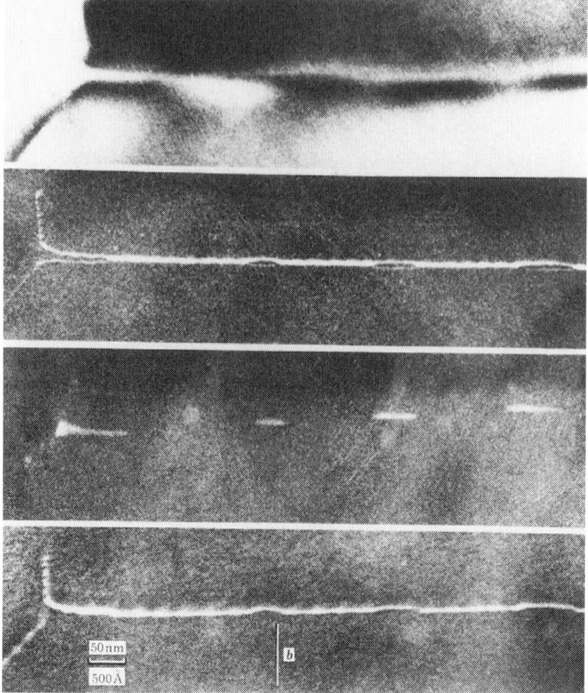

Figure 2.3 Weak beam images of a dissociated edge dislocation in silicon with constricted segments. From the top: a strong beam image in which the constricted segments are masked; a weak beam image showing both partial dislocations; a weak beam image for which $g \cdot b = g \cdot b \times u = 0$ for the total dislocation, and the stacking fault of width approximately 8 nm is observed; a weak beam image in which one partial dislocation and the stacking fault are out of contrast.

Cockayne *et al* [1] established a simple formula for converting the separation of image peaks into dissociation widths. This formula corrects for the fact that image peaks are predicted to lie at different distances from dislocation cores, because the scattering geometry for the two partials is not identical. This is a systematic correction, which becomes increasingly important for smaller partial separations. When the separations of the dislocation lines are large ($\geqslant 15$ nm), the differences between measured image peak separations and dislocation line separations can be ignored, but as the separations become smaller (towards 2 nm) the correction becomes increasingly important. In addition there are inaccuracies in the determination of the dislocation core positions due to shifts in the image peak position with dislocation depth and foil thickness. These inaccuracies reduce as $|s|$ is increased, but they are

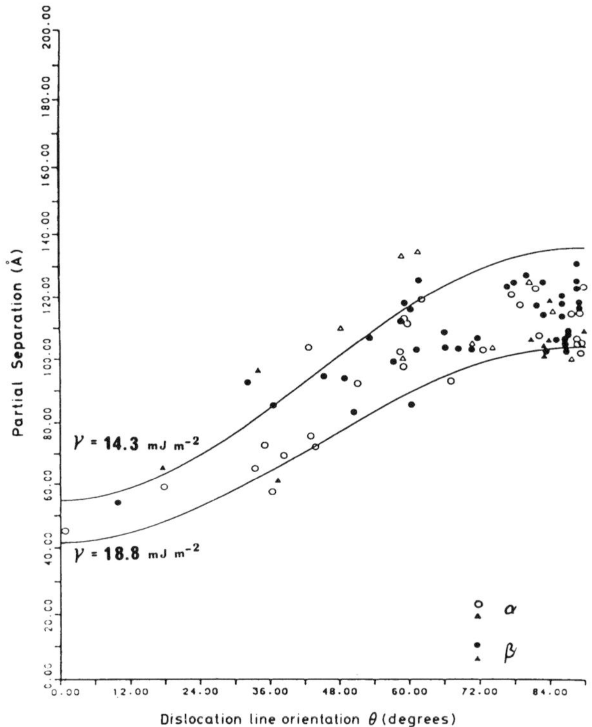

Figure 2.4 The variation in measured dissociation width of Shockley partials as a function of dislocation line orientation in CdTe (α and β dislocations). The theoretical curves based on elasticity theory for two values of stacking-fault energy are shown [63].

typically 1 nm. They can be minimized by averaging results over a range of foil thicknesses or dislocation depths, or by using a convergent electron beam (see e.g. [7]).

When a range of measured dissociation widths Δ is displayed as a function of dislocation orientation (θ) (e.g. as in figure 2.4), either isotropic or anisotropic elasticity theory can be used to fit to the data, with the stacking-fault energy (γ) as a fitting parameter. In this way accurate determinations of γ for an enormous number of materials have been obtained (e.g. in f.c.c. materials [8], materials with the diamond structure [6, 9, 10], diamond [11] and III–V and II–VI semiconductors [12, 13]). For materials where suitable thin crystals can be obtained, the weak beam technique is the most accurate method for determining γ.

For the partial dislocation separations smaller than 3 nm, there is the possibility that overlap of dislocation cores might produce deviations from separations predicted by continuum elasticity theory using unextended cores. As an example, the experimental values for Cu (in which the partial cores for the screw orientation are $\simeq 2$ nm apart) were considered in detail by several authors (see e.g. [14, 15]), and the influence of core models on the value of γ obtained was investigated [16].

The technique becomes unreliable for measuring dissociation widths smaller than ≈ 2 nm, because the uncertainty in the relationship between the image peak positions and the dislocation cores becomes greater than the dissociation width. Pirouz *et al* [11] approached this limit in a study of dislocation dissociation in diamond, where they established that the dislocations are dissociated and so determined the stacking-fault energy.

Weak beam imaging has been used to study the change in dissociation width with various experimental parameters. Hazzledine and Karnthaler [17] investigated the influence of surface relaxation on dissociation width; Saka *et al* [18, 19] studied the dependence of γ on specimen temperature; Barreteau and Loiseau [20] investigated antiphase boundaries (APBs) in Fe + 27 at.% Al as a function of temperature; Nakada and Imura [21] looked at the effects of dopants and deformation temperature on γ in GaAs; Stoltz and Vander Sande [22] investigated the effect of nitrogen segregation on γ in Fe–Ni–Cr–Mn steels.

Dislocation dissociation in various minerals has been investigated (e.g. quartz [23] and olivine [24]), as well as in numerous metals and alloys.

Extensive weak beam studies have been made of dislocation dissociation and dislocation structures in semiconductors. The demonstration that dislocations in semiconductors are dissociated [6] helped to resolve the debate about whether these dislocations are of the glide or shuffle set (see Hirsch [25]), and the question of whether or not they remain dissociated during glide was resolved by capturing weak beam video images of dissociated dislocations gliding in CdS [26].

The determination of fault energies from dislocation dissociation widths can be extended to situations where there is multiple dissociation, as in the fourfold dissociation in DO3-ordered alloys. In this system, ordering of both the first and second nearest-neighbours occurs. The unit dislocation then consists of four identical dislocations connected by APBs. The outer two pairs have an APB with energy γ_1, and the inner pair have an APB with energy γ_2 (figure 2.5). The separations of these dislocations are a complex function of γ_1 and γ_2, but a plot of measured separations as a function of line orientation can be used to give a best fit to theoretical curves based on elasticity theory, using these two energies as fitting parameters (figure 2.5). In this example, from Fe + 26 at.% Al, the dislocation separations are so large ($\simeq 200$ Å) that image peak separations can be taken as dislocation separations. In figure 2.5 the absence of data points in the range $105°–156°$ agrees well with the predicted unstable orientations predicted by Head [27].

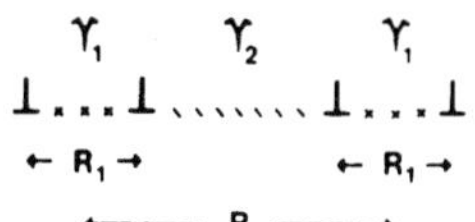

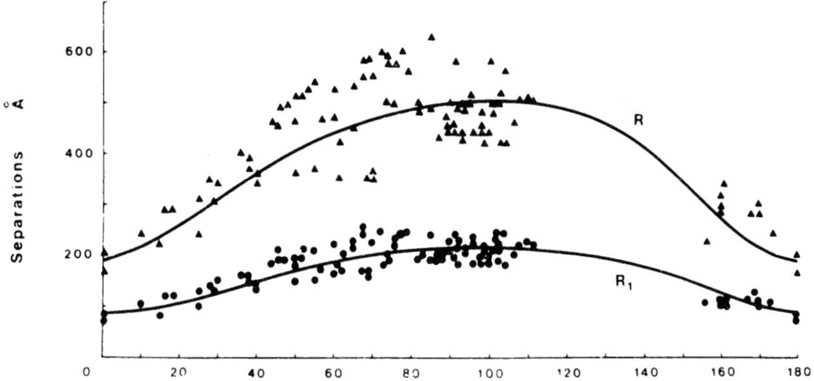

Figure 2.5 Separations of fourfold dissociation of DO3-type superlattice dislocations in Fe + 26 at.% Al as a function of line orientation, as measured from weak beam images [28].

2.3 COMPLEX DISLOCATION GEOMETRIES

The weak beam method is particularly useful for revealing the geometry of complex dislocation interactions and geometries, such as the delineation of dislocation nodes, of high density dislocation networks and of interface dislocations. An interesting case is the study by Crawford *et al* [28] of the fourfold dissociation in the DO3-ordered alloy Fe + 26 at.% Al, referred to in the previous section. In this material, although theory predicts certain dislocation line directions to be unstable, due to the anisotropy of the material, strong beam images show dislocations taking on these 'forbidden' line directions. Weak beam images (e.g. figure 2.6) show that the dislocations can achieve these unstable directions by being composed of segments of line lying in stable directions. The figure shows a clear example of a defect geometry that would be impossible to investigate by either strong beam or HREM techniques. A second example is the complex geometry of dislocations in silicon (figure 2.3), where the strong beam image gives no indication of the complex dissociation geometry revealed by the weak beam image.

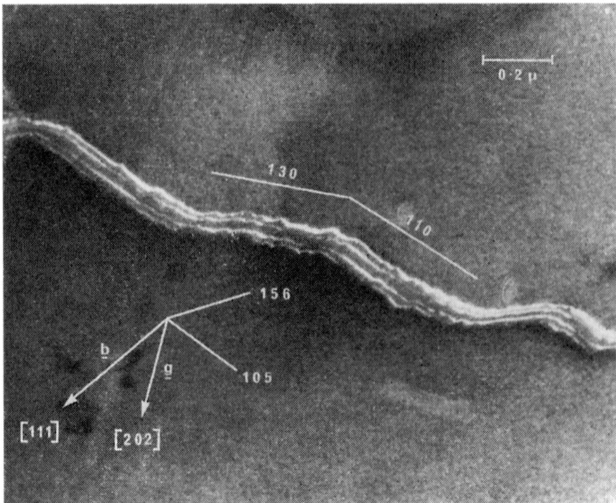

Figure 2.6 Weak beam image showing fourfold dissociation of superlattice dislocations in a DO3-ordered Fe + 26 at.% Al alloy. The average orientations of the dislocation lines (as indicated by straight lines) are unstable for a material of this anisotropy, but are seen to be composed of very fine zig-zags [28].

2.4   SMALL REGIONS OF STACKING FAULTS

The conditions for obtaining stacking-fault contrast using large $|s|$ are the same as for strong beam images, *viz.* $g \cdot R \neq$ integer where R is the displacement vector of the fault. Images taken under these conditions can be used to provide useful information about defects.

For inclined faults, the depth periodicity of stacking-fault fringes for large $|s|$ is approximately $|s|^{-1}$. Consequently for large $|s|$, faults lying parallel to the foil surface show contrast which is very sensitive to the fault depth, while inclined faults can appear as narrow closely spaced fringes (figure 2.7). The direction of these narrowly spaced fringes (which is parallel to lines of constant fault depth) can be used to determine the fault plane.

For a small region of fault bounded by a partial dislocation line, the image when $s \approx 0$ is often dominated by the influence of the strain field of the partial. The use of large $|s|$ can provide a means of observing the fault image because of the diminished width of the image of the surrounding partial dislocation [29] (e.g. figure 2.3). Small loops can be investigated in this way to determine whether or not they are faulted. Other studies which have used weak beam stacking-fault images to determine defect geometries include the investigation of geometries of dissociated dislocations in silicon [6], the construction of a detailed model

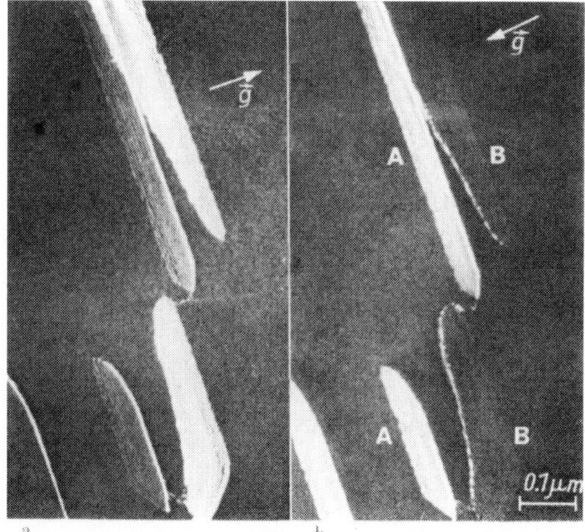

Figure 2.7 Weak beam images of inclined extrinsic stacking faults in silicon, taken with $\pm g$.

of point-defect clusters in ion-irradiated materials (e.g. [30]), and the study of {311} and {111} rodlike defects in self-ion-irradiated silicon [31, 32].

2.5   INVISIBILITY CRITERIA

Many of the standard rules which are used to analyse defects in strong beam images apply equally well to weak beam images. For example the normal $g \cdot b = 0$ and $g \cdot b \times u = 0$ invisibility criteria for dislocations apply to weak beam images (as figure 2.3 demonstrates), as do the inside–outside contrast rules used to determine the directions of Burgers vectors for small loops. However, it is important to appreciate that for dissociated dislocations, while $g \cdot b = 0$ might apply to the total dislocation, the individual partials might have $g \cdot b \neq 0$. This point applies equally well to strong beam images, and accounts for residual contrast often seen in those images when invisibility is expected, but in weak beam images the narrowness of the images of the individual partial dislocations allows their invisibility, and hence their Burgers vectors, to be studied individually. Figure 2.3 clearly demonstrates this, with lengths of undissociated dislocation showing no contrast, and dissociated lengths showing contrast from the partials and the associated stacking fault. At the same time, anisotropy can cause a breakdown of the invisibility criteria, just as it does in strong beam images.

2.6   STRAINED INCLUSIONS

Inclusions (e.g. precipitates) can strain a matrix, giving rise to complex strong beam images which can be analysed by image simulation. If the maximum strain is small, then no part of the lattice might be sufficiently misoriented to give rise to strong contrast under weak beam imaging conditions. This is the situation for quantum dots in low-strain systems (Liao and Cockayne, to be published). However, in other systems there is sufficient strain to allow weak beam studies to be carried out. For example Guyot and Renault [33] investigated spherical precipitates and Melander [34] cuboidal precipitates.

2.7   MISFIT DISLOCATIONS

In strained layer semiconductors with low strain, pre-existing threading dislocations can relieve strain by extending in the interface (the 'Matthews mechanism') under the influence of the misfit strain. This occurs when the strained layer thickness exceeds a 'critical thickness'. Weak beam images have shown that in many of the semiconductors used in heterostructure device material the threading dislocations are dissociated. Zou and Cockayne [35] investigated the resulting misfit dislocations in InGaAs/GaAs, combining high resolution imaging and weak beam microscopy. They showed that the misfit dislocations are also dissociated. Since the dissociation plane is not generally the interface plane, it is apparent that models of misfit dislocations which have them lying at the interface cannot be correct at the atomic level. Zou and his colleagues showed that the position of the partial dislocations with respect to the interface depends upon the nature of the strain (tensile or compressive), and can be explained in terms of elasticity theory. This has important implications for quantum well devices since the dissociation structure can often be comparable to the well width.

Strain relief occurs by the glide of threading dislocations under the influence of the misfit strain, but opposed by the dislocation line tension. However, since these threading dislocations are generally dissociated, and since the resolved stress on each partial is different, there is a different critical thickness for each partial. In a series of papers, Zou and Cockayne [36] investigated the effect of this dislocation dissociation on the Matthews mechanism, taking into account the different forces on each partial. Two cases arise—when the resolved stress on the leading partial exceeds that on the trailing partial and vice versa (figure 2.8). In the latter case the trailing partial forces the leading partial to glide ahead of it, and the dissociation width is similar to the static condition; in the former case, the leading partial glides ahead of the trailing partial, with the possibility of the leading partial gliding to infinity before the critical thickness for the trailing partial is reached. This leads to the formation of extensive stacking faults, with subsequent deleterious effects on electronic properties.

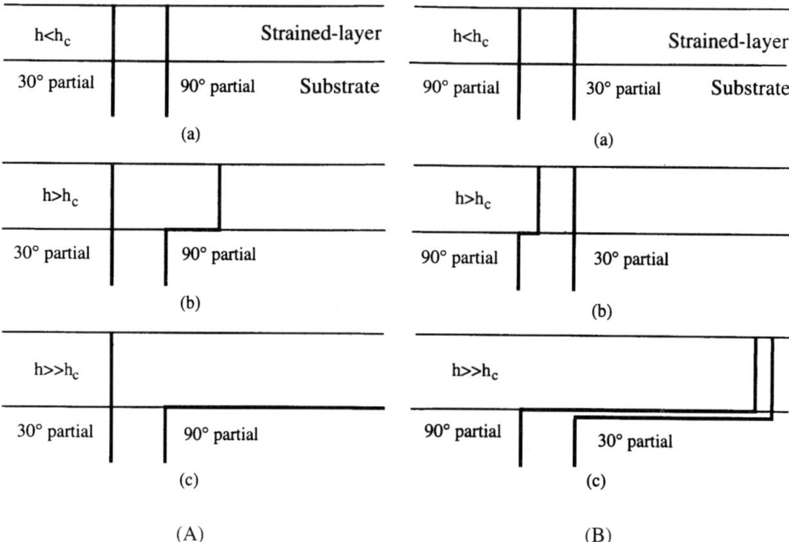

Figure 2.8 Schematic diagrams of the misfit dislocation generation mechanism where the force on the 90° partial is greater than that on the 30° partial. Two cases are shown (90° leading (A) and 90° trailing (B)), each for increasing epilayer layer thickness.

2.8   WEAK BEAM DEFECT ANALYSIS

The use of invisibility criteria and the imaging of small areas of stacking fault (as in figure 2.3) allows small faulted and unfaulted loops to be studied, and their interaction mechanisms to be unravelled. Because the images are well defined, and closely map the projection of dislocation cores, weak beam stereo images are particularly useful for unravelling the three-dimensional arrangements of networks and defects. For example Breen [37] has studied the arrangement of defects in $GaAs/In_xGa_{1-x}/As$ interfaces by this means, while many authors have investigated dense regions of small loops. Kumar and Hemker [38] have characterized dislocation structures in small angle grain boundaries in the $L1_2$ intermetallic alloy Ni_3Ge, where they showed the closely spaced superdislocations to be dissociated. Other applications include studies of interactions of dislocations with precipitates [39], the effect of temperature on dislocation structures [40], phase transformations [41] and the radiation-induced formation of helical dislocations [42].

The weak beam technique has particular value in determining Burgers vectors where there are several different kinds of defect in regions of high defect density. In an investigation of defects caused by 1 MeV silicon implantation in silicon, followed by annealing, Chou *et al* [43] determined the line directions

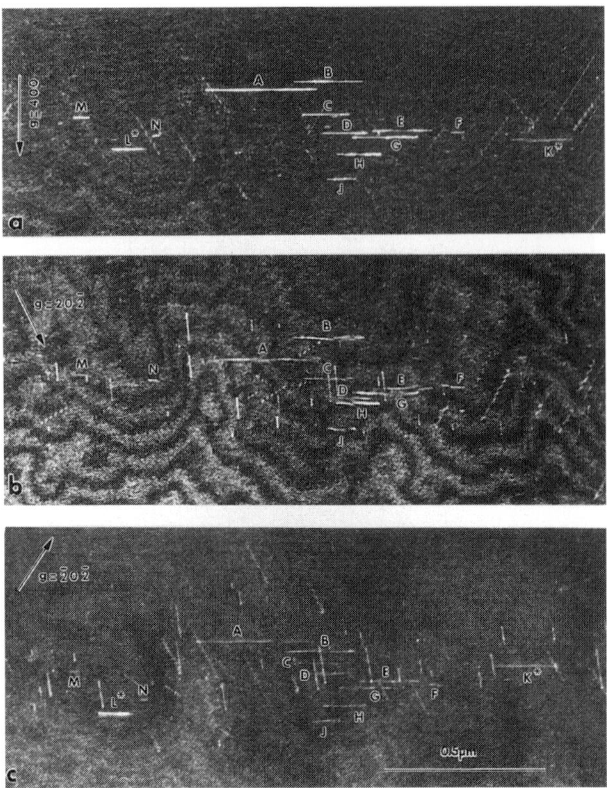

Figure 2.9 Weak beam images for ion-irradiated silicon, taken in three different reflections, showing two kinds of rodlike defect (see the text).

and Burgers vectors of the resulting rodlike defects by a systematic analysis of weak beam images taken in a number of reflections. Figure 2.9 shows a set of images showing these defects, with individual defects labelled. Particular defects can be seen to disappear in different reflections; by considering the change in visibility of individual defects in the various diffraction vectors, it was shown that, contrary to the accepted view, not all are {311} rodlike defects (i.e. long lengths of extended defect lying on {311} with displacement vectors $a\{611\}/k$ where $25 < k < 31$); rather, approximately 14% of them have {111} habit planes. Rodlike defects can be an important source of interstitials leading to transient enhanced diffusion [44] at elevated temperatures. Since the stability of the {111} defects is likely to be different from the stability of {311} defects, and since the number of interstitials trapped in the two defects will be different, this observation necessitates a recalculation of their influence on enhanced diffusion. Energy minimization calculations, carried out on interstitial

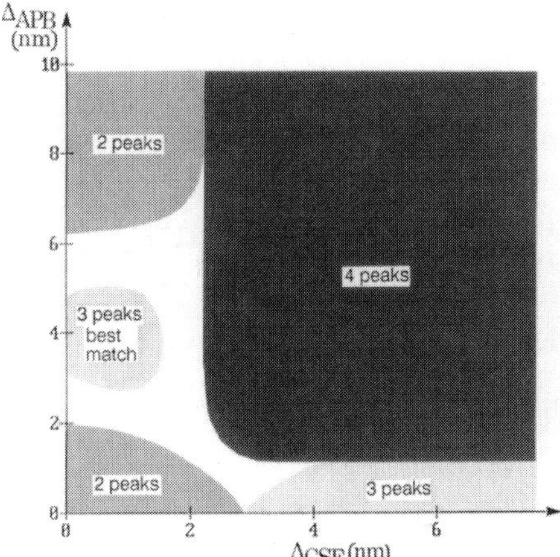

Figure 2.10 Diagram illustrating the number of image peaks of simulated edge dislocations as a function of the separation of APB and complex stacking-fault (CSF) widths, for a fourfold dissociated superlattice dislocation (for details see [53]).

atomic configurations suggest a displacement vector $a\langle 111\rangle/10$ for the {111} defects. Molecular dynamics studies are being made to further refine the models of both the {311} and the {111} rodlike defects, and to provide an understanding of their early stages of formation.

2.9 SUPERLATTICE DISLOCATIONS

The weak beam technique has been used in numerous studies of superlattice dislocations in ordered alloys (e.g. [45–51]), and in particular to determine the APB energies in these technologically important materials. Because of the large Burgers vectors in these materials, many of these studies use reflections for which $g \cdot b > 2$. It has to be remembered that this breaks the conditions for weak beam images (*viz.* $g \cdot b \leqslant 2$), and this can give rise to multiple image peaks even from an undissociated dislocation. When the partials are widely separated, this causes few difficulties, but in many of the systems studied (e.g. Ni_3Al), this is not the case, and as a consequence the images cannot be easily interpreted.

One approach to overcoming this problem is to simulate images for a range of defect parameters, and seek the best fit between simulated and experimental images. The technique of Head [3] for simulating images of inclined dislocations is ideal for this purpose, provided that the grey scale and dimensions are adjusted to handle the high contrast and narrow images. Schaublin and Stadelmann [52] have developed a form of this simulation programme specifically for weak beam imaging.

Korner *et al* [53] turned the variable number of image peaks into an advantage by simulating images for the relevant experimental parameters (i.e. dislocation separations), and plotting the number of image peaks observed as a function of these parameters (figure 2.10). By establishing these plots for several different diffracting vectors, the corresponding experimental images were used to define a range of APB energies for which the experimental images are consistent with the simulated images.

2.10   PLANAR DEFECTS

An area of analysis not yet fully investigated is the use of weak beam images for studying planar defects (e.g. stacking faults and heterostructure interfaces). Although these defects are often modelled as planar defects with no structural width normal to the defect plane, in fact they can have a finite width in two senses: (1) structurally (e.g. the extrinsic stacking fault extends over two {111} planes), and (2) extended variations in the electron scattering potential (e.g. the scattering potential on either side of an intrinsic fault is disturbed by the presence of the fault). The sensitivity of weak beam images to local scattering power can be used to probe this structure. The principles are most easily explored with reference to the intrinsic and extrinsic stacking faults and the amplitude–phase diagram. Considering first the extrinsic fault, the amplitude–phase diagram for this fault, taken under weak beam conditions, is shown in figure 2.11. The image amplitude for the fault, averaged over fault depth in the foil, is related to the distance between the centres of the two circles. The radius of the large circle is $1/(2\pi |s|)$, and the length of the small arc corresponds to the width of the fault in the direction of the incident beam (assuming the column approximation). For $|s| \approx 2 \times 10^{-2}$ Å^{-1}, the circle has radius 8 Å, which is comparable to the fault arc length of d_{111}. If the sign of $(\mathbf{g} \cdot \mathbf{R})s$ is reversed (by reversing $\mathbf{g}$ or changing the sign of s), the amplitude–phase diagram changes as in figure 2.11(*b*) to figure 2.11(*c*). The asymmetry of figures 2.11(*b*) and 2.11(*c*) results in an appreciable change in the contrast of extrinsic stacking-fault fringes in images with the opposite sign of $(\mathbf{g} \cdot \mathbf{R})s$. An example is seen for extrinsic faults in silicon in figure 2.7. This effect was reported in early weak beam studies of extrinsic faults [54–56], and was explained in the above terms by several authors [57–59]. As the inclination of the fault to the incident beam direction increases, this asymmetry increases, because the fault thickness in the direction of the beam increases with inclination. Cockayne *et al* [59] investigated this effect quantitatively, and showed good agreement with theory.

Turning now to the intrinsic fault, on the above argument the stacking-fault contrast would be independent of the sign of $(\mathbf{g} \cdot \mathbf{R})s$ because the intrinsic fault has no width (being on one {111} plane). However, experimental studies

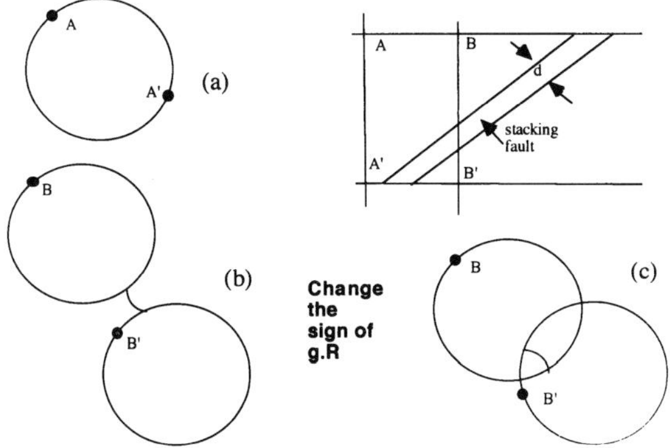

Figure 2.11 Amplitude–phase diagram for scattering from an extrinsic stacking fault. The sign of $g \cdot R$ is changed between (*b*) and (*c*).

(e.g. [60]) indicated that there is an asymmetry. Multislice calculations [60] also showed this asymmetry, with the contrast highly dependent on the fault inclination and the value of s. The image asymmetry arises because of the effect of the fault on the scattering potential on either side of the fault, i.e. the disturbed scattering potential has a finite width even though the geometrical fault is planar.

This sensitivity of weak beam images to the structure of planar defects remains to be fully exploited.

2.11   CONCLUSIONS

As the examples described above show, the weak beam technique is now well established as a basic tool for the study of crystalline defects. Its central principle—the sensitivity of weakly excited reflections to local structure—can be extended to other studies (e.g. the imaging of atomic steps on surfaces [61]).

The technique derived its name from the fact that the image taken with large $|s|$ is generally very weak in intensity, compared with strong beam images, even though the image contrast is high. Because of the weak image intensity, correcting focus and astigmatism is difficult, and techniques were developed to overcome this problem. Transmission microscopes were fitted with two beam-tilt channels, so that focus and astigmatism could be corrected on the bright-field image, with the second channel reserved for weak beam imaging. However, the development of image intensifiers and CCD cameras has made it possible to see and focus weak beam images without difficulty. Astigmatism can be corrected directly on the weak beam image, and the image can be collected digitally, to give access to on-line analysis. The image is further improved by the use of field emission guns and energy filtering systems to minimize the effects of chromatic aberration. Weak beam imaging with bright guns and image intensification is now as straightforward as normal bright-field imaging, and, for this reason, many experimentalists use the technique for routine imaging.

REFERENCES

[1]   Cockayne D J H, Ray I L F and Whelan M J 1968 *Phil. Mag.* **20** 1265–70
[2]   Whelan M J 1978 *Diffraction and Imaging Techniques in Materials Science* vol 1, ed S Amelinckx, R Gevers and J van Landuyt (Amsterdam: North-Holland) pp 43–106
[3]   Head A K 1969 *Aust. J. Phys.* **22** 43–106
[4]   Cockayne D J H 1972 *Z. Naturf.* a **27** 452–60
[5]   Cockayne D J H 1978 *Diffraction and Imaging Techniques in Materials Science* vol 1, ed S Amelinckx, R Gevers and J van Landuyt (Amsterdam: North-Holland) pp 153–183
[6]   Ray I L F and Cockayne D J H 1971 *Proc. R. Soc.* A **325** 543–54
[7]   Holmes S M, Cockayne D J H and Ray I L F 1974 *Proc. 8th Int. Congr. on EM (Canberra)* ed J V Sanders and D J Goodchild (Canberra: Australian Academy of Science) pp 290–1
[8]   Jenkins M L 1972 *Phil. Mag.* **26** 747–51
[9]   Haussermann F and Schaumburg H 1973 *Phil. Mag.* **27** 745–51
[10]   Ray I L F and Cockayne D J H 1973 *J. Microsc.* **98** 170–3
[11]   Pirouz P, Cockayne D J H, Sumida S, Hirsch P B and Lang A 1983 *Proc. R. Soc.* A **386** 241–9
[12]   Cockayne D J H and Hons 1979 *J. Physique Coll.* **40** C6 11–18
[13]   Gomez A M and Hirsch P B 1978 *Phil. Mag.* **37** 733–7
[14]   Cockayne D J H, Jenkins M L and Ray I L F 1971 *Phil. Mag.* **24** 1383–92
[15]   Stobbs W M and Sworn C H 1971 *Phil. Mag.* **24** 1365
[16]   Cockayne D J H and Vitek V 1974 *Phys. Status Solidi* b **65** 751–64
[17]   Hazzledine P M and Karnthaler H P 1975 *Phil. Mag.* **32** 81–97
[18]   Saka H and Iwata T 1978 *Phil. Mag.* **37** 291–6
[19]   Saka H and Suekiu Y 1978 *Phil. Mag.* **37** 273–89
[20]   Barreteau C and Loiseau A 1994 *J. Physique Coll.* **4** C3 65–73
[21]   Nakada Y and Imura T 1987 *Phys. Status Solidi* a **102** 625–32
[22]   Stoltz R E and Vander Sande J B 1980 *Metall. Trans.* A **11** 1033–7
[23]   Trepied L and Doukan J C 1978 *J. Mater. Sci.* **13** 492–8
[24]   Vander Sande J B and Kohlestedt D L 1976 *Phil. Mag.* **34** 653–8
[25]   Hirsch P B 1980 *J. Microsc.* **118** 3–12
[26]   Cockayne D J H, Hons A and Spence J C H 1980 *Phil. Mag.* **42** 773–81
[27]   Head A K 1967 *Phys. Status Solidi* **19** 185–192
[28]   Crawford R C, Ray I L F and Cockayne D J H 1973 *Phil. Mag.* **27** 1–7
[29]   Cockayne D J H, Jenkins M L, Ray I L F and Whelan M J 1971 *Proc. 25th Anniv. Meeting EMAG (Cambridge)* ed W C Nixon (London: Institute of Physics) pp 108–9

[30] Jenkins M L 1974 *Phil. Mag.* **29** 813–20
[31] Chou C T and Cockayne D J H 1995 *Phys. Rev.* B **53** 17 223–30
[32] Chou C T, Cockayne, D J H, Krinhoj P and Jagadish C 1997 *Proc. 1996 Conf. on Optoelectronic and Microelectronic Materials and Devices (Canberra)* ed C Jagadish (Piscataway, NJ: IEEE) pp 305–8
[33] Guyot P and Renault A 1976 *J. Microsc. Spectrosc. Electron.* **1** 23–31
[34] Melander A 1976 *Phys. Status Solidi* **33** 255–64
[35] Zou J and Cockayne D J H 1994 *J. Appl. Phys.* **75** 7317–22; 1995 *J. Appl. Phys.* **77** 2448–53
[36] Zou J and Cockayne D J H 1993 *J. Appl. Phys.* **74** 925–30; 1994 *Phys. Rev.* B **49** 8086–95
[37] Breen K R 1992 *J. Electron. Mater.* **21** 401–18
[38] Kumar M and Hemker K J 1997 *Phil. Mag. Lett.* **76** 399–407
[39] Summerfelt S R and Carter C B 1992 *Acta Metall. Mater.* **40** 2805–12
[40] Yoshida M and Takasugi T 1992 *Phil. Mag.* **66** 89–102
[41] Hitzenberger C and Karnthaler H P 1991 *Phil. Mag.* **64** 151–63
[42] Oliver J and Veyssiere P 1991 *Phil. Mag. Lett.* **63** 141–51
[43] Chou C T, Cockayne D J H, Zou J, Kringhoj P and Jagadish C 1995 *Phys. Rev.* B **52** 12 223–30
[44] Eaglesham D J, Stolk P A, Gossmann H-J and Poate J M 1994 *Appl. Phys. Lett.* **65** 2305–7
[45] Ray I L F, Crawford R C and Cockayne 1990 *Phil. Mag.* **21** 1027–32
[46] Korner A 1995 *Mater. Sci. Eng.* A **192/193** 262–7
[47] Korner A 1992 *Phil. Mag.* **66** 103–18
[48] Zhang S and Milligan W W 1995 *Phil. Mag.* **71** 523–36
[49] Veyssiere P and Noebe R 1992 *Phil. Mag.* **65** 1–13
[50] Baluc N H and Karnthaler 1991 *Phil. Mag.* **64** 137–50
[51] Baluc N and Schaublin R 1991 *Phil. Mag. Lett.* **64** 327–34
[52] Schaublin R and Stadelmann P 1993 *Mater. Sci. Eng.* A **164** 373–8
[53] Korner A, Cockayne D J H and Sun Y Q 1993 *Phil. Mag.* **68** 993–1001
[54] Cullis A G and Booker G R 1972 *Proc. 5th Eur. Congr. on Electron Microscopy (Manchester)* (London: Institute of Physics) pp 532–3
[55] Ferreira Lima C A and Howie A 1976 *Phil. Mag.* **34** 1057–71
[56] Salisbury I G and Loretto M H 1979 *Phil. Mag.* **39** 317–323
[57] Foll H, Carter C B and Wilkens M 1980 *Phys. Status Solidi* **58** 393–407
[58] Wilkens M, Foll H and Carter C B 1982 *Phys. Status Solidi* **73** K15–9
[59] Cockayne D J H, Pirouz P, Liu Z, Anstis G R and Karnthaler P 1984 *Phys. Status Solidi* **82** 425–39
[60] Wilson A and Cockayne D J H 1985 *Phil. Mag.* **51** 341–54
[61] Kambe K and Lehmpfuhl G 1975 *Optik* **42** 187–94
[62] Hirsch P B, Howie A and Whelan M J 1960 *Phil. Trans. R. Soc.* A **252** 499
[63] Lu G and Cockayne D J H 1986 *Phil. Mag.* **53** 307–20

3

TWO-BEAM AND n-BEAM DIFFRACTION

Alec F Moodie

Jeffreys has remarked that approximation is one of the subtlest of all the skills in mathematical physics as well as one of the most rewarding. There can be few more striking examples of the truth of this dictum than that afforded by the work of Hirsch, Howie and Whelan in adapting the two-beam approximation to describe the imaging of defect structures in the electron microscope [1].

Their techniques are now well known and widely practised by all of those in solid state physics, chemistry, metallurgy and materials science who are concerned with defects; their results are of central importance in all of those fields, their analysis has been extended [2] and elaborated [3] but not superseded, yet the depth of their insight in the selection of two-level formalism as the appropriate language in which to describe the scattering of fast electrons by the defect crystal structure passes almost unremarked so inevitable does it now seem.

However, at an accelerating voltage of, say, 200 keV, and for all its subtlety, the two-beam approximation affords rather a poor description of scattering from a perfect crystal. That it nevertheless, and in various guises, finds wide application stems ultimately from its structure which differs fundamentally from that deriving from conventional perturbation theory.

This can be conveniently illustrated in terms of scattering [4]. The exact forward scattering solution can be written down as a series, each term of which can be represented as a graph in momentum space, the vertices being points in the reciprocal lattice weighted by excitation errors and the edges representing structure amplitudes. The order of the graph, counting loops in the conventional way, is the order of the interaction and the contribution of each graph to the wave function is calculated by means of simple rules.

In a conventional perturbation approximation, graphs up to a given order are summed, for instance up to one in the kinematical approximation, while in the finite beam approximations graphs between a finite number of reciprocal lattice points are summed to infinite order. The simplest of such approximations involves two beams, the forward scattered beam and one diffracted beam g with excitation error ζ_g. Two structure amplitudes are necessarily involved, namely $V(g)$ and $V(\bar{g})$ which couple the diffracted beam to the central beam. As Cowley and Moodie have shown [5] this infinite series of partial sums can be evaluated by means of combinatorial algebra to obtain the standard two-beam solution. While this is not a particularly simple way to derive the solution it illustrates graphically how the form is built up from notional elementary scattering events and emphasizes that all orders of interaction are included in the approximation. It is essentially this factor which, taken in conjunction with the numerical values of the wavelength λ at the accelerating voltage W generating a velocity v, and the interaction constant σ [6],

$$\sigma = \frac{\pi}{W\lambda} \frac{2}{1 + \left(1 - v^2/c^2\right)^{1/2}} \tag{3.1}$$

that guarantees the robustness of the form, though not the numerical accuracy, of the two-beam approximation when appropriate angles of incidence are chosen. It is, of course the form, and only the form, of the two-beam solution that is central in the Hirsch–Howie–Whelan analysis and this constitutes one of the subtlest aspects of their approximations.

In order to obtain a necessary condition that a scattering configuration should be reducible to two-beam form it is convenient to write the solution in terms of the scattering matrix S and the structure matrix M_0 so that

$$S = \exp\{iM_0 z\} \tag{3.2}$$

with z the thickness of the crystal.

Purely for compactness the projection approximation has been made, that is M_0 has been assumed to be independent of z.

Then,

$$\langle g|S|0 \rangle = \sum_{j=1}^{n} \langle g|P_j|0 \rangle \exp\{i\mu_j z\} \tag{3.3}$$

where the μ_i are the eigenvalues of M_0 and the P_j are Dirac projectors given by

$$P_j = \prod_{\substack{l=1 \\ l \neq j}}^{n} (M_0 - \mu_l E)/(\mu_j - \mu_l) \tag{3.4}$$

with E the unit matrix.

The necessary but not sufficient condition that a particular beam $\langle g|$ should be of two-beam form is then that all but two of the P_j should be zero, that is [7] that $\langle g|M_0^2|0\rangle / V(g)$ should be equal to the sum of two of the eigenvalues of M_0. With M_0 Hermitian this quantity must be real and so there is no orientation at which the non-centrosymmetric three-beam approximation will reduce to two beam, since $2\pi\zeta_g + \sigma V(h)V(h-g)/V(g)$ will in general be complex. There are however seven-beam configurations which reduce to two-beam form in 15 of the non-centrosymmetric space groups [8].

Many reductions are known in the centrosymmetric space groups [9] and these have been classified in terms of the underlying group structure in zone axis orientations [10]. The three-beam approximation falls into a category of its own here. Niehrs [11] introduced the concept of reduction in solving the three-beam zone axis case.

At the very outset however, in calculating $\langle g|M_0^2|0\rangle / V(g)$ the term $\sigma V(h)V(h-g)/V(g)$ emerges in the equation

$$2\pi\zeta_g + \sigma \frac{V(h)V(h-g)}{V(g)} = \mu_1 + \mu_2. \tag{3.5}$$

There are three possible terms of this type in the three-beam approximation, namely

$$\frac{V(h)V(h-g)}{V(g)} \equiv G \tag{3.6}$$

$$\frac{V(g)V(h-g)}{V(h)} \equiv H \tag{3.7}$$

$$\frac{V(g)V(h)}{V(h-g)} \equiv C. \tag{3.8}$$

These are precisely the structure invariants introduced by Hauptman in his study of the phase problem of structure analysis [12]. The phase associated with those quantities is independent of the origin chosen to define atomic positions in the crystal and depends only on the crystal structure. In the Hauptman–Karle formulation of structure analysis by direct methods, probabilities are assigned to the phases of initially a few of these structure invariants, broadly on the basis of the distribution of kinematical intensities with scattering angle, and an iterative process is then invoked to enlarge the set. The emergence of structure invariants as a direct consequence of the physics of dynamical scattering suggests that a method may exist for recovering all of the parameters required for structure analysis in centrosymmetric crystals by direct measurement wherever the three-beam approximation has adequate validity, that is, for inverting the three-beam approximation. This proves to be correct [13] and the key factor lies in determining the loci of excitation error along which the intensity distributions in a convergent beam electron diffraction (CBED) pattern have two-beam form for all thicknesses. $|V(g)|$, $|V(h)|$, $|V(h-g)|$ and the phase of the structure invariant can then be determined directly from three distances measured on the pattern.

The symmetries used in this analysis are the point symmetry of inversion in conjunction with the symmetry SU(3) of the group of the dynamical differential equations and the antisymmetry of the appropriately phase shifted two-beam form. In fact the sub-algebras of su(3) that lie in su(2), the algebra of two-level systems, are identified and the resulting restrictions are sufficient to invert the system. Alternatively, instead of the symmetry of the Lie group of the defining differential equation, the symmetry of the characteristic equation, that of S_3, isomorphous with that of D_{3h}, can be invoked by means of the standard substitution [14].

The phase of the structure invariant is determined by the sign of the displacement from the Bragg angle of the centre of any one of the five two-beam loci. Since this depends only on the form of the intensity distribution and not on the thickness or the structure amplitude, and since there are five independent checks, signs may be determined fairly readily, at least in unit cells of moderate size.

Limitations of the three-beam approximation guarantee that the apparent structure amplitudes will in fact be pseudo-potentials generated by the remainder of the beams. While, with the correct phases, these may be sufficiently close to the true structure amplitudes to establish a workable trial structure, it is of some interest to determine whether the analysis can be extended to higher approximations. If precisely the same arguments are used, then in the four-beam approximation the five two-beam loci degenerate to a point and the method fails. An argument used by Cayley in 1849 shows very simply how this comes about. A line in three-dimensional space has four degrees of freedom and the cubic has an equal number of disposable parameters; therefore it must be possible to rule at least one line on a cubic surface but not in general on a surface of higher degree. In fact, as Cayley and Salmon pointed out, many lines can be ruled on the general cubic surface. In the present context the lines are the loci and the cubic surface one of the dispersion surfaces in the Bloch wave formulations.

This approach must therefore be modified and indeed the obvious symmetry generated by interchanging couplers and diffracted beams which merely leads to useful redundancy in the three-beam approximation can be invoked. Development however is not yet sufficiently far advanced to be useful.

The non-centrosymmetric three-beam approximation can now be analysed by starting from the inversion of the centrosymmetric case and noting that the transformation from one symmetry to the other is continuous. Two intersecting loci pass through each diffracted beam when a centre of inversion is present and these are properly described as one curve, namely a degenerate hyperbola. As the inversion condition is relaxed so that the antisymmetric parts of the structure amplitudes increase smoothly in magnitude from zero the degeneracy is lifted and the loci become the asymptotes of a hyperbola which traces these angles of incidence that generate eigenvalues of a specified form. The two-beam condition is, in fact, never attained but is approached asymptotically. While this approach serves to clarify the connection between centrosymmetric and non-centrosymmetric cases it is not yet clear whether it can find any application in structure analysis.

In all of the cases considered so far the reduction to two-beam diffraction has been exact, but this can frequently be relaxed usefully in a number of directions. One of these has already been referred to, namely the representation of the effects of other diffracted beams by means of a pseudo-potential. The importance and effectiveness of this approximation, particularly in small unit cells, was recognized by Bethe [15] in this first theoretical description of dynamical electron diffraction. Gjønnes [16] has generalized this approach and set it on more secure foundations.

Because of such effects numerical values in the two-beam approximation are rarely more accurate than five per cent and frequently much worse, but the form of the solution is unaffected, and here the subtlety of the Hirsch–Howie–Whelan approximation is seen to great effect. In addition systematic weak beams must always be present but these obey the same rules for zero contrast as the satisfied beam, and again the approximation is effective.

Yet another n-beam effect can be incorporated in a modified two-beam description and this can be seen in striking form as a Borrmann effect in convergent beam diffraction. When a notional two-beam condition is set up in convergent beam diffraction a quantitative estimate of the validity of the approximation in maintaining the form can be obtained by measuring the symmetry of the intense diffracted beam about the Bragg position and noting effects due to any of the non-systematic interactions that may happen to cross the disc. For structures with a unit cell of moderate dimensions the departure from symmetry will commonly be small. The forward scattered beam, in contrast, will almost certainly be strongly asymmetric, a typical Borrmann effect which can be well described, by analogy with x-ray diffraction, by a phenomenological absorption coefficient. It can, however, prove instructive to determine the orientation with some accuracy and then carry through a number of n-beam purely elastic calculations, say, with 50, 200 and 500 beams. A likely outcome is that the intense diffracted beam will retain its symmetry to good approximation but that the forward scattered beam will develop an asymmetry which can be pronounced, though it will never quite equal that in the experiment. Clearly yet another n-beam effect from the point of view of defect analysis can be well described phenomenologically, this time as an apparent absorption.

A convenient method for obtaining perturbation descriptions of such quasi-two-beam effects, as well as the extension of, say, zone axis reductions to cover convergent beam discs is to use the absolutely convergent expansion of the exponent $\exp\{A + B\}$ where A and B are non-commuting matrices.

The expansion is due to Zassenhaus and is of the form

$$\exp\{A + B\} = \exp A \exp B \exp P_1 \exp P_2 \ldots \tag{3.9}$$

where $P_1 = -1/2[A, B]$, and succeeding terms involve nested commutators of increasing order guaranteed to be smaller than the preceding term.

Both theory and experiment therefore establish clearly that, at the commonly used wavelengths, n-beam diffraction is always significant, but that, nevertheless, at carefully chosen angles of incidence a phenomenological formulation of two-beam diffraction is very likely to give a good account of experimental results, and that any technique which depends on the form, as distinct from the parameters of the intensity distribution, will prove eminently successful. This, appreciated at the outset by Hirsch, Howie and Whelan, has proved to be one of the factors which has allowed them to establish a technique of great generality and power with an economy of means which has opened applications to many of us who might otherwise have been excluded.

Given now the wide applicability of two-beam forms the formidable task of incorporating a description of the defect lattice in such a form remains. The means by which this was achieved and the numerical techniques devised to elucidate defect structures are by now well known to all who work in this extremely wide field, as the most cursory inspection of the extensive literature will attest.

The effectiveness and subtlety of the procedures which have been devised can, perhaps, be best appreciated by specific consideration of the Howie–Whelan equations. Head [17] in an illuminating analysis wrote these equations as

$$\frac{\mathrm{d}T}{\mathrm{d}z} = (\mathrm{i} - \alpha)S$$

$$\frac{\mathrm{d}S}{\mathrm{d}z} = (\mathrm{i} - \alpha)T + \mathrm{i}B(z)S \tag{3.10}$$

where T is the amplitude of the forward scattered beam, S the amplitude of the diffracted beam and α the phenomenological absorption coefficient, and $B(z)$ describes the local distortion of the crystal down the column.

Putting $R = S/T$ these equations become

$$\frac{\mathrm{d}R}{\mathrm{d}z} = (\mathrm{i} - \alpha)\left(1 - R^2\right) + \mathrm{i}BR. \tag{3.11}$$

This equation cannot be solved in terms of elementary functions but Head shows how to choose $B(z)$ so that the equation does have an explicit solution, general when $\alpha = 0$ and particular otherwise. Apart from providing a valuable check on numerical procedures this work throws light on the nature of the solution, and, indeed, in a sense establishes a general approach to the solution.

Equations of the Howie–Whelan form have, of course, been widely studied in other branches of physics and chemistry but only a limited number of approximate solutions are known. Something of their group structure can be learned by adapting a technique due to Fried [18], notionally applicable to n-beam scattering. In the case of purely elastic scattering, that is $\alpha = 0$, defining $\left(\begin{smallmatrix} T \\ S \end{smallmatrix}\right) \equiv |b\rangle$, and putting $|b\rangle = |q\rangle \exp\left\{\mathrm{i}\frac{1}{2}\int_0^z B(z)\,\mathrm{d}z\right\}$ the Howie–Whelan equations can be written

$$\frac{\mathrm{d}|q\rangle}{\mathrm{d}z} = \mathrm{i}\begin{pmatrix} -\frac{1}{2}B(z) & \sigma V(\bar{g}) \\ \sigma V(g) & \frac{1}{2}B(z) \end{pmatrix}|q\rangle \equiv \mathrm{i}\left(a\sigma_1 + b\sigma_2 - \tfrac{1}{2}B(z)\sigma_3\right) \tag{3.12}$$

where a and b are the real and imaginary parts of the structure amplitude $V(g)$, admissably a function of z, and the σ_j are the Pauli matrices, the conventional basis for problems with the symmetry of SU(2). The unitary transformation U characterized by the matrix

$$\frac{1}{\sqrt{2}}\begin{pmatrix} 1 & \sin\phi + \mathrm{i}\cos\phi \\ -\sin\phi + \mathrm{i}\cos\phi & 1 \end{pmatrix}$$

with ϕ the phase of $V(g)$, eliminates the diagonal terms so that the equation becomes, with $|p\rangle \equiv \left(\begin{smallmatrix} p_0 \\ p_1 \end{smallmatrix}\right) = U|q\rangle$,

$$\frac{\mathrm{d}|p\rangle}{\mathrm{d}z} = \mathrm{i}\begin{pmatrix} 0 & a+\pi B(z)\sin\phi - \mathrm{i}(b-\pi B(z)\cos\phi) \\ a+\pi B(z)\sin\phi + \mathrm{i}(b-\pi B(z)\cos\phi) & 0 \end{pmatrix}|p\rangle \tag{3.13}$$

$$\equiv \mathrm{i}(\sigma_1 E_1 + \sigma_2 E_2)|p\rangle, \quad \text{defining } E_1 \text{ and } E_2 \tag{3.14}$$

a form that draws attention to the need for polarity calibrated image rotation when carrying out defect analysis in a non-centrosymmetric crystal [19].

Writing $E_\pm = E_1 \pm \mathrm{i}E_2$, and $E_+ E_- = E_1^2 + E_2^2$, the equation becomes

$$\frac{\mathrm{d}^2 p_0}{\mathrm{d}z^2} + \left(E_1^2 + E_2^2\right)p_0 - \left\{\frac{\mathrm{d}E_-/\mathrm{d}z}{E_-}\right\}\frac{\mathrm{d}p_0}{\mathrm{d}z} = 0 \tag{3.15}$$

i.e.

$$\frac{\mathrm{d}^2 p_0}{\mathrm{d}z^2} + \left(E_1^2 + E_2^2\right)p_0 - \left\{\frac{\mathrm{d}}{\mathrm{d}z}(\ln E_-)\right\}\frac{\mathrm{d}p_0}{\mathrm{d}z} = 0. \tag{3.16}$$

Once again this equation is suggestive of forms for $B(z)$ for which solutions in terms of elementary functions can be constructed. Since a standard substitution reduces this equation to the form

$$\frac{\mathrm{d}^2 y}{\mathrm{d}z^2} = f(z)y \tag{3.17}$$

the analysis can be carried further in terms, at least, of classifying the solutions. This has not yet been carried through, and indeed, in view of the breadth of the treatment of Hirsch, Howie and Whelan, is unlikely to prove to be of practical value.

This returns us to our starting point, namely that the insights which led Hirsch, Howie and Whelan to constructing their theory have been equalled by their skills in providing solutions. What would appear to be, of necessity, complex and intractable, they have made transparent.

REFERENCES

[1]　Hirsch P, Howie A, Nicholson R B, Pashley D W and Whelan M J 1965 *Electron Microsocopy of Thin Crystals* (London: Butterworth)

[2]　Cockayne D J H, Ray I L F and Whelan M J 1969 *Phil. Mag.* **20** 1265

[3]　Howie A and Basinski Z S 1968 *Phil. Mag.* **17** 1039

[4]　Gjønnes J and Moodie A F 1965 *Acta. Crystallogr.* **19** 65

[5]　Cowley J M and Moodie A F 1957 *Acta. Crystallogr.* **10** 609

[6]　Dawson B, Goodman P, Johnson A W S, Lynch D F and Moodie A F 1974 *Acta. Crystallogr.* A **30** 279

[7]　Moodie A F and Fehlmann M 1993 *Acta. Crystallogr.* A **49** 376

[8]　Moodie A F and Whitfield H J 1994 *Acta. Crystallogr.* A **50** 730

[9]　Blume J 1966 *Z. Phys.* **191** 248

[10]　Fukahara A 1966 *J. Phys. Soc. Japan* **21** 2645
　　　Kogiso M and Takahashi H 1977 *J. Phys. Soc. Japan* **42** 223

[11]　Niehrs H 1961 *Int. Conf. on Magnetism and Crystallography (Kyoto, 1961)* paper 232

[12]　Hauptman H A 1972 *Crystal Structure Determination—the Role of the Cosine Seminvariants* (London: Plenum)

[13]　Moodie A F, Etheridge J and Humphreys C J 1996 *Acta. Crystallogr.* A **52** 596

[14]　King R B 1996 *Beyond the Quartic Equation* (Boston, MA: Birkhäuser)

[15]　Bethe H A 1928 *Ann. Phys., Lpz.* **87** 55

[16]　Gjønnes J 1962 *Acta. Crystallogr.* **15** 703

[17]　Head A K 1981 *Phil. Mag.* A **44** 827

[18]　Fried H M 1988 *J. Math. Phys.* **30** 1161

[19]　Brown P D, Loginov Y Y, Stobbs W M and Humphreys C J 1995 *Phil. Mag.* **72** 39

4

PSEUDO-ABERRATION-FREE FOCUSING IMAGING METHOD FOR ATOMIC RESOLUTION ELECTRON MICROSCOPY OF CRYSTALS

H Hashimoto

4.1   INTRODUCTION

With the improvement of the performance of transmission electron microscopes (TEMs) and the development of theories of image formation, conventional TEMs have now attained atomic resolution, are widely used in many fields of science and technology and are producing many useful results. This tendency has been accelerated by improvements in several functions of the microscopes, e.g. development of imaging lenses with very small aberrations, increases in the coherency of the illuminating electron beams, improvements in elemental analysis by detection of characteristic x-rays and by energy loss spectroscopy and design of specimen chambers with various types of specimen treatment device involving heating, cooling and reactive and non-reactive gas atmospheres.

With the present types of microscope the aim is to improve the resolution by elevating the accelerating voltage, say to 1 MeV or more, and to reduce the spherical aberration to a minimum towards zero. However, the general applicability of such an instrument has been questioned. For example, at very high voltage, the excessive radiation damage will significantly modify the detailed structure of the specimens or of the crystal defects within a few seconds [1] and a zero spherical aberration lens will not give any contrast on the images of a weak phase object (WPO).

The medium high voltage electron microscopes such as 300–400 kV have now reached a point to point resolution of 0.18–0.16 nm in the conventional definition for WPOs [2], which is smaller than the atomic distances in metal crystals, and are producing electron microscope images which are useful for materials science. Though the electron microscope images give the impression that the contrast represents the structure of the specimen faithfully, they must not be understood intuitively, but interpreted by considering the modes of scattering of electrons and the process of image formation by electron lenses with aberrations. The images of thin films are formed by the interference of waves scattered from the specimens in both elastic and inelastic modes. Moreover, the electrons illuminating the specimens are not completely coherent but partially coherent. Thus the image contrast becomes complicated especially when spherical and chromatic aberrations disturb the image quality. By the use of field emission and point cathode electron guns, the coherence and current density of illuminating electron waves have been improved recently and waves scattered through large angles can contribute to the image contrast.

The observed electron microscope images with atomic resolution, however, change with small changes of focus of the objective lens. The change of image contrast by defocusing of the objective lens is well described by the phase contrast transfer function (CTF) [3]. The phase of the waves leaving the specimen at some scattering angle is shifted by the spherical aberration of the imaging lens but it can be compensated by underfocusing. Thus the phase angle of the CTF is given by the difference of the terms due to spherical aberration, $\pi/2\lambda(C_s\alpha^4)$, and defocus, $\pi/2\lambda(2\Delta f\alpha^2)$, where λ, C_s, α and Δf are the wavelength, spherical aberration coefficient, electron scattering angle and defocus.

Scherzer [3] proposed the optimum defocus value of $\Delta f = (C_s\lambda)^{1/2}$ for the WPO in which the CTF can be approximated by its imaginary part and a wide flat region appears close to the optical axis (small scattering angle) in the oscillating CTF. The first zero of the oscillating CTF appears at the scattering angle of $\alpha_m = 1.41(\Delta f/C_s)^{1/2}$. Eliminating the oscillating region of the CTF (for waves scattered through angles larger than α_m) by the objective aperture, the resolution limit becomes $\delta = C_s^{1/4}\lambda^{3/4}$, which is now called conventionally the theoretical resolution limit. By adopting the allowance of the defocus which gives a nearly flat region in the CTF, a higher resolution limit is obtainable and now the value $\delta = RC_s^{1/4}\lambda^{3/4}$, where $R = 0.65$–0.7, is generally used as the conventional resolution limit.

In general the lattice images of thin crystalline films a few tens of nanometres in thickness are formed by the interference of many waves diffracted under dynamical conditions. Since these diffracted waves have suffered different phase shifts the Scherzer focus may not be applicable for obtaining the correct images with the contrast similar to the intensity distribution of the electron waves at the bottom surface of the crystal.

In this paper, the nature of the image contrast of crystals formed by partially coherent waves under the spherical aberration-free focus (AFF) condition, which enables images of crystals to be obtained while avoiding the effect of spherical aberration, is discussed. This is followed by a discussion of the pseudo-AFF (ψ-AFF) which can be applied to obtain the images of two or more crystal regions with different lattice spacings such as those on either side of an interface.

4.2 IMAGES OF CRYSTALS FORMED BY PARTIALLY COHERENT ILLUMINATION

The behaviour of electron waves which are passing through the conventional image forming lens system after Bragg reflection in a crystal can be well treated by the theory which takes into account the partial coherence of the waves, formulated by Ishizuka [4]. When the imaging waves are partially coherent, the intensity of the electron waves at the image plane is expressed as the summation of all interfering pairs of individual Bragg reflected waves, i.e.

$$I(r) = \sum \sum T(g_2 g_1)\phi_{g2}\phi_{g1} \exp 2\pi \mathrm{i}(g_2 - g_1)r \tag{4.1}$$

where ϕ_{g2} and ϕ_{g1} are the wave amplitudes reflected from the planes g_2 and g_1 and leaving the bottom face of the crystal. $T(g_2 g_1)$ is the transmission cross coefficient (TCC) and represents the order of interference between two image forming waves ϕ_{g2} and ϕ_{g1}. It acts to reduce the degree of interference and is expressed as

$$T(g_1 g_2) = \exp[-\mathrm{i}\{\gamma(g_2) - \gamma(g_1)\}]E_s E_f E_x P \tag{4.2}$$

where

$$\gamma(g) = (\pi/2)\left(C_s \lambda^3 g^4 - 2\Delta f \lambda g^2\right) \tag{4.3}$$

and

$$g = 1/d = \alpha/\lambda.$$

E_s is a term due to beam divergence, E_f a term due to chromatic defocus value Δ, $E_x = 1$, $P = 1$. When the illumination is coherent, the product of E_s, E_f, E_x and P becomes unity.

Figure 4.1 shows the values of TCC at 400 kV for a JEOL-4000EX microscope, plotted against chromatic defocus values for the diffracted waves from the lattice planes, 000, 111, 200, of an Au crystal in the [110] orientation.

The intensities of Bragg reflected waves are generally strong and the interference effects between two Bragg reflected waves at the same angles relative to the optical axis contribute strongly to the image contrast as can be seen in figure 4.1.

For a WPO, the intensity of the scattered waves is very small and thus the interference of scattered waves with the primary wave is predominant and that between scattered waves becomes negligibly small. Thus by setting $g_2 \rightarrow g$, $g_1 \rightarrow 0$

$$T(g, 0) = \exp(-\mathrm{i}\gamma(g))E_s E_f. \tag{4.4}$$

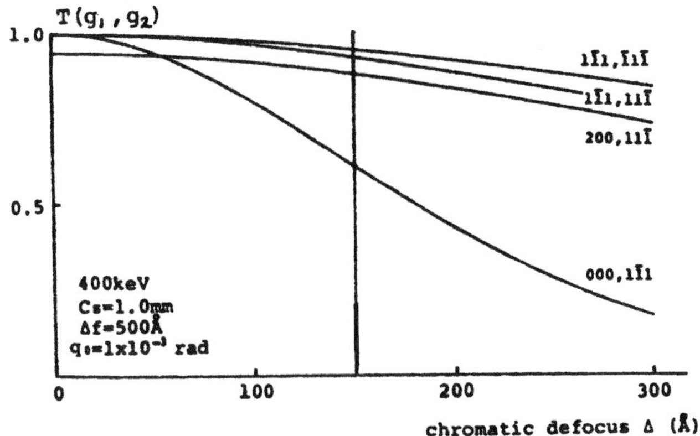

Figure 4.1 Variation of TCC at 400 kV against chromatic defocus values Δ, for $C_s = 1$ mm, $\Delta f = 50$ nm, $q = 1 \times 10^{-3}$ rad.

E_s and E_f are envelope functions to the phase CTF $\exp(-i\gamma(g))$ and decrease with increasing the scattering angle of the electrons.

The above theoretical predictions can be illustrated in the following observations shown in figure 4.2. Figure 4.2(*a*) shows the electron diffraction pattern of an MgO crystal in the [110] orientation together with the objective aperture, which includes the waves diffracted from 111, 002 and 220; (*b*) is the electron microscope image of the MgO crystal formed with the condition shown in (*a*): the spacing of 220 ($d = 0.149$ nm) is clearly resolved; (*c*) is the Fourier transform of (*b*), which appears as if the waves reflected from 440 ($d = 0.074$ nm) have contributed to the image. This does not mean that the image in (*b*) represents the real structure of the MgO crystal with a resolution of 0.074 nm. The fine structure of 0.074 nm, which corresponds to 440, was produced by the interference of two waves, which are diffracted by the two planes 220 and $\bar{2}\bar{2}0$.

The spot corresponding to 331 was produced by the interference of 111 and $\bar{2}\bar{2}0$ and not by the interference of the waves from 000 and 331. This is due to the interference between the diffracted waves with high intensity.

Figures 4.2(*d*), (*e*) and (*f*) are from amorphous silicon film prepared by deposition onto rocksalt. The amorphous diffraction pattern was selected by the objective lens aperture as shown in (*d*) and produced the image shown in (*e*). The Fourier transform of (*e*) is shown in (*f*) which suggests that the structure with the space frequencies corresponding to the dark concentric rings in (*f*) is not recorded in (*e*). By comparing the figures of the (*a*), (*b*), (*c*) series with those of the (*d*), (*e*), (*f*) series, it is seen clearly that even if the same size of aperture is used, the recorded image quality is very different and both images need detailed interpretation in terms of the image formation process.

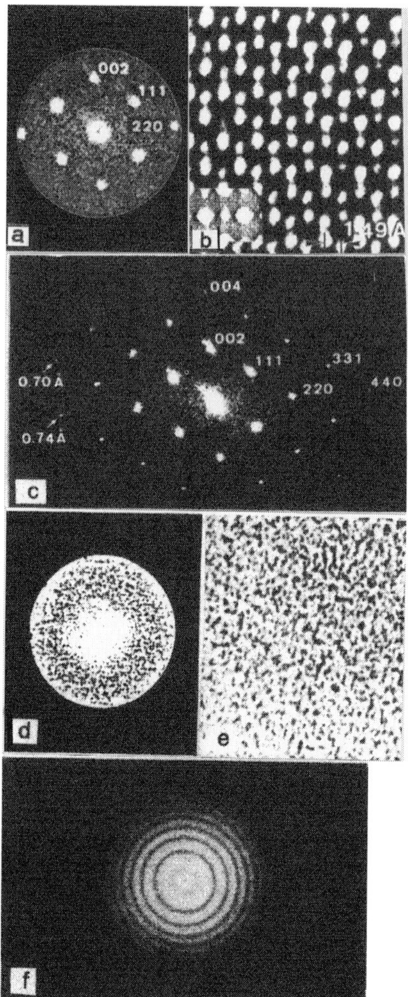

Figure 4.2 Relation of electron diffraction patterns, apertures, images and Fourier transforms of images for an MgO crystal and for amorphous Si.

4.3   IMAGES UNDER ABERRATION-FREE FOCUSING CONDITIONS

4.3.1   Principle of the AFF condition

The principle of the AFF condition [5] is described here by using some simple methods and illustrations.

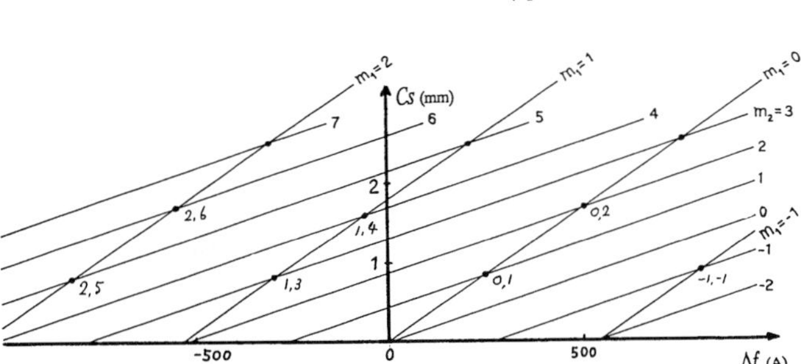

Figure 4.3 C_s–Δf diagram for the AFF condition. For {200} this is defined by the $m_1 = 0, \pm 1, \pm 2, \ldots$ line system, and for {220} by the $m_2 = 0, \pm 1, \pm 2, \ldots$ line system for an MgO crystal in the [001] projection. The crossing points of these two line systems indicated by dots give the AFF condition for both {200} and {220}.

The phase CTF, which is expressed as $\exp(-i\gamma(g))$, should satisfy

$$\exp(-i\gamma(g)) = 1 \tag{4.5}$$

by eliminating the effect of spherical aberration and defocus, where $\gamma = (\pi/2)(C_s\lambda^3/d^4 - 2\Delta f\lambda/d^2)$. This is expressed by assuming $\gamma(g_1) = 2m_1\pi$, which gives

$$\left(\lambda^3/4d_1^4\right)C_s - \left(\lambda/2d_1^2\right)\Delta f = m_1 \tag{4.6}$$

for the reciprocal lattice vector g_1 with spacing d_1, where $m_1 = 0, \pm 1, \pm 2, \ldots$. Equation (4.6) is expressed as a set of lines in Δf–C_s diagrams as shown in figure 4.3 for a $\langle 200 \rangle$ projection of an MgO crystal at 400 kV ($\lambda = 0.016\,43$ Å), $d_1 = 0.2106$ nm. The combination of the values of C_s and Δf at any point on these lines becomes the AFF condition for four {200} planes.

In these conditions Δf is given from equation (4.6) as

$$\Delta f = m_1\left(2d_1^2/\lambda\right) + \left(C_s\lambda^2/2d_1^2\right) \tag{4.7}$$

and is shown schematically in figure 4.4 [6]. The diffracted waves from the lattice planes with the spacing d_1 will not be focused on the Gaussian image plane but on a plane shifted by an amount $C_s\lambda^2/2d_1^2$. The intensity distribution at the bottom surface can be obtained by adjusting the focus to this plane, i.e. $\Delta f = C_s\lambda^2/2d_1^2$. However, on both sides of this image plane, Fourier images [7] with the period of $m_1(2d_1^2/\lambda)$, where $m_1 = 0, \pm 1, \pm 2, \ldots$, are also produced. Since the Fourier images have the same intensity distribution as the original

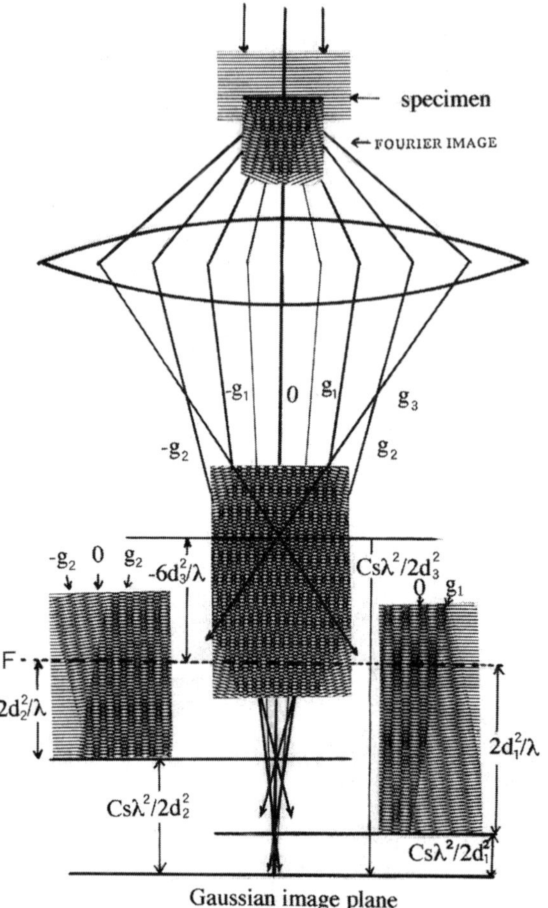

Figure 4.4 Ray diagram for the imaging condition, showing the relationship between the AFF condition and Fourier images. The intensity distribution at the bottom surface of the crystal specimen formed by the interference of the diffracted waves from the lattice planes with lattice spacing d_1 is projected on a plane shifted from the Gaussian image plane by $C_s\lambda^2/2d_1^2$. On both sides of this plane, Fourier images with period of $m(2d_1^2/\lambda)$ are produced [7]. (After figure 2 in [14], courtesy of Elsevier Science.)

image formed at $(C_s\lambda^2/2d_1^2)$, the defocus Δf in equation (4.7) gives the focus which is equivalent to an exact focus with a zero-aberration lens.

A similar relationship holds for another lattice plane, say (220) of MgO ($d_2 = 0.1480$ nm), which is also shown in figure 4.3 as $m_2 = 0, \pm1, \pm2, \ldots$. The

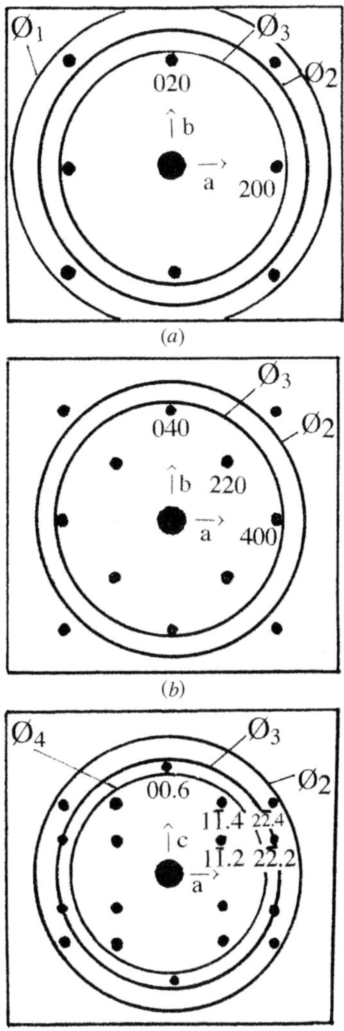

Figure 4.5 Relationship between various sizes of apertures and electron diffraction patterns from (*a*) MgO, (*b*) MgAl$_2$O$_4$, (*c*) Al$_2$O$_3$, $\phi_1 = 7$ nm^{-1} ($d = 0.14$ nm), $\phi_2 = 6$ nm^{-1} ($d = 0.17$ nm), $\phi_3 = 5$ nm^{-1} ($d = 0.2$ nm), $\phi_4 = 4$ nm^{-1} ($d = 0.25$ nm). (From figure 3 in [14], courtesy Elsevier Science.)

AFF conditions for two kinds of plane {200} and {220} are given as the crossing points of these two line systems. They are indicated with the combination of numbers (m_1, m_2) and are listed in table 4.1.

Table 4.1 AFF conditions for MgO ($d_1 = 0.2106$ nm) (200) ($d_2 = 0.1480$ nm) (220).

m_1	m_2	C_s (mm)	Δf (nm)
2	5	0.811 16	−83.212
1	3	0.832 37	−28.604
0	1	0.853 59	26.003
−1	−1	0.874 80	80.611
2	6	1.664 75	−57.209
1	4	1.685 96	−2.601
0	2	1.707 18	52.007
6	23	9.262 19	−41.609

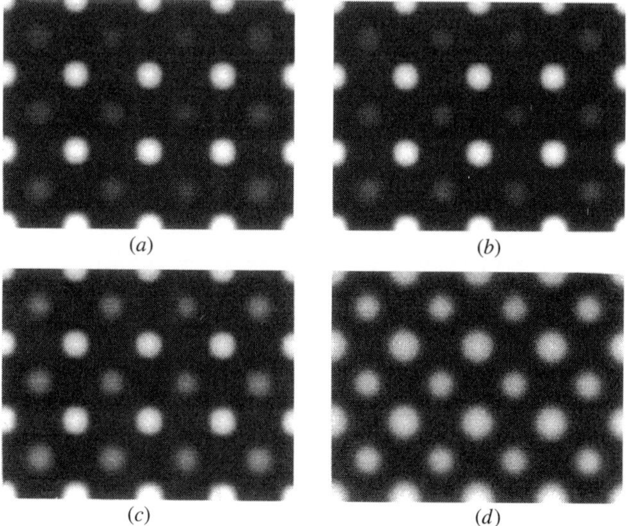

Figure 4.6 MgO in the [001] orientation, 400 kV, $t = 50$ nm, $\phi_1 = 7$ nm^{-1}. (*a*) intensity at bottom surface, (*b*) image intensity at AFF ($C_s = 0.8535$ mm, $\Delta f = 26.00$ nm), (*c*) AFF, $q = 0.1$ mrad, $\Delta = 1.0$ nm, (*d*) AFF, $q = 1$ mrad, $\Delta = 15$ nm.

These AFF conditions correspond to the coincidence of two Fourier images of $m_1(2d_1^2/\lambda)$ and $m_2(2d_2^2/\lambda)$ formed on both sides of the image planes ($C_s\lambda^2/2d_1^2$) and ($C_s\lambda^2/2d_2^2$) respectively, as shown in figure 4.4. For a crystal with a simple structure in a symmetry orientation, more than two Fourier images from more than two planes coincide at a plane as shown by plane g_3 in figure 4.4.

For an MgO crystal in a [001] orientation, four reflections from {200} and four from {220} are in the same phase under these conditions as shown in figure 4.3

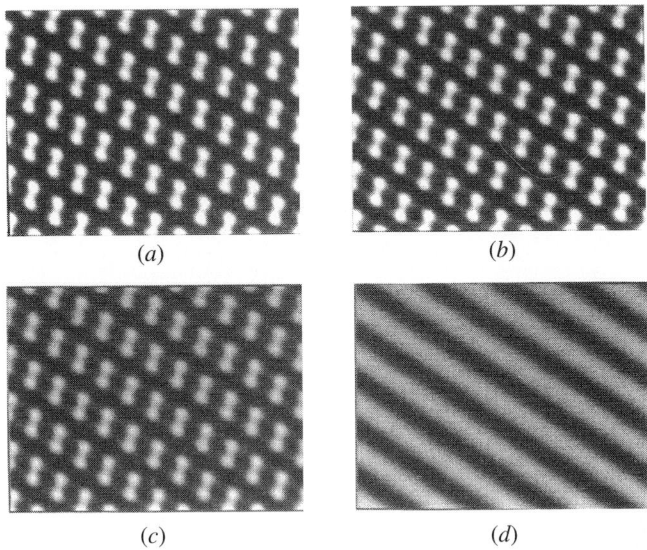

Figure 4.7 Effect of partial coherence of illuminating electrons on the image contrast of an Al_2O_3 crystal for an AFF condition $C_s = 1.15$ mm, $\Delta f = -134.59$ nm and $\phi_2 = 6$ nm^{-1}. (*a*) Intensity at the bottom surface, (*b*) AFF, $q = 0$ mrad, $\Delta = 0$ nm, (*c*) AFF, $q = 0.1$ mrad, $\Delta = 1$ nm, (*d*) AFF, $q = 1$ mrad, $\Delta = 10$ nm.

and table 4.1, and produce an image which has the same intensity distribution of electrons as that at the bottom surface of the crystal, if the objective aperture selects only these nine diffraction spots to contribute to the image contrast as indicated in figure 4.5(*a*). Most of the conventional atomic resolution electron microscopes have resolution limits around $\delta = 0.14$–0.25 nm, which correspond to the apertures of $\phi_1 = 7$–4 nm^{-1} and thus, in the general case, it is possible to use the waves which are diffracted from two kinds of lattice plane with low index for producing the atomic structure images.

As can be seen in figure 4.6(*b*), images of atoms in an MgO crystal in a [001] orientation formed under AFF conditions ($C_s = 0.8535$ mm, $\Delta f = 26.00$ nm) with an aperture $\phi_1 = 7$ nm^{-1} ($d = 0.143$ nm) are similar to the intensity distribution of electrons at the bottom face shown in (*a*). Figures 4.6(*c*) and (*d*) are the images formed by partially coherent waves, i.e. beam divergences are $q = 0.1$ and 1 mrad, and chromatic defocus values are $\Delta = 1.0$ and 15 nm, respectively. Though the relative intensity at the atomic positions in (*d*) is not as strong compared with the background as in (*b*) and (*c*), the position of the atoms can be identified. Figure 4.7 for an Al_2O_3 crystal in the [11.0] projection shows images for large values of C_s and Δf ($C_s = 1.15$ mm, $\Delta f = -134.6$ nm), for various values of q and Δ. In this case, the contrast in figures 4.7(*c*) and (*d*) is so

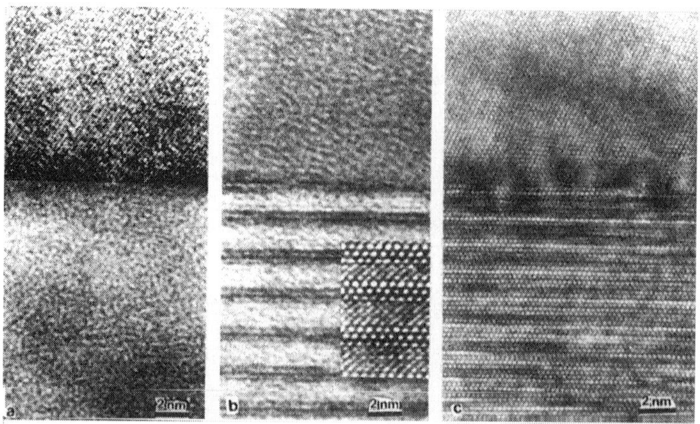

Figure 4.8 Images of interfaces in different focus in (a) and (b), Ag (top), BiPbSrCaO (bottom) [8]; (c) interface in composite TiAl (top), Ti$_2$AlC (bottom) [9].

small that the image intensity has been printed using contrast enhancement, in which the lowest contrast is set to zero and the highest set to unity. In practice, however, the contrast can hardly be seen in figures 4.7(c) and (d), even though the coherence of the illuminating electrons is relatively high, particularly for case (c), and C_s and Δf correspond to the AFF condition.

Though there are many AFF conditions, not all of them are useful if the partial coherence of electron waves is taken into account.

4.3.2   Pseudo-aberration-free focusing

The atomic structure images of crystals on both sides of an interface are not so easy to record at the same time, because, even if there is no misorientation, there is some difference of lattice spacings of the crystal. Figures 4.8(a) and (b) are the cross-sectional images of an interface between Ag and high T_c superconductor BiPbSrCaO with different focus [8]. Only the image of Ag(111) $d = 0.24$ nm is seen in figure 4.8(a) and that of BiPbSrCaO $d = 2.3$ nm (001) is seen in figure 4.8(b). By comparing with the inset in figure 4.8(b), which is the well adjusted image, it is seen that there is difficulty in taking high quality structure images of the materials on both sides of the interface at the same time. Figure 4.8(c) is the cross-sectional image of a composite of Ti$_2$AlC in a TiAl matrix [9]. Atomic structure images of Ti$_2$AlC, TiAl and their interface are clearly recorded and contribute to the understanding of the high ductility of this composite material.

Figure 4.9 shows the images of two interfaces and three crystal lattice images of Al$_2$O$_3$, MgAl$_2$O$_4$ and MgO [10]. The images of three crystals are seen clearly with high contrast in spite of the lattice spacings having large differences. Even

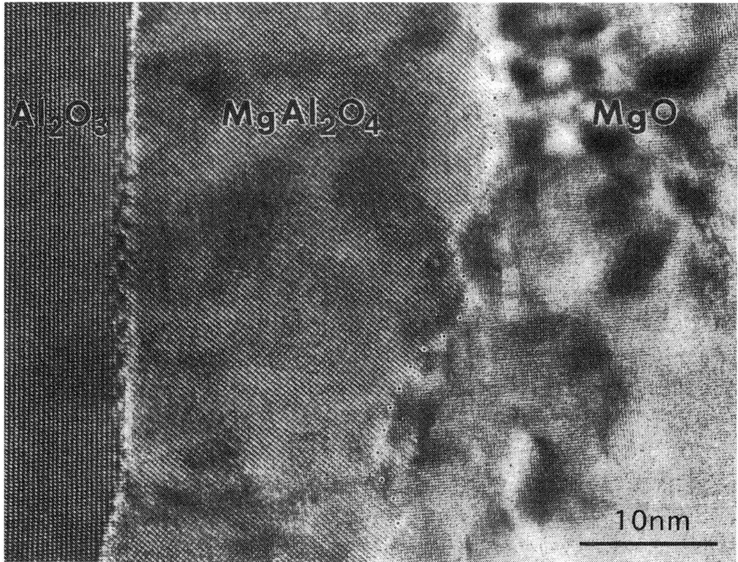

Figure 4.9 Interface formed by the reaction between MgO and Al$_2$O$_3$ [10].

though these lattice spacings are within the resolution limit (0.18 nm) and the thickness of the specimen is same, the focusing of the lens plays an important role in obtaining the images which have similar intensities at the bottom surfaces of the crystals. This condition is referred to as the ψ-AFF condition. This will now be discussed, using the example of Al$_2$O$_3$, MgAl$_2$O$_4$ and MgO, shown in figure 4.9.

In order to apply the ψ-AFF condition to this system, the AFF diagrams for MgO, MgAl$_2$O$_4$ and Al$_2$O$_3$ are drawn as shown in figure 4.10. Their corresponding AFF conditions are in tables 4.1–4.3. They are superimposed as shown in figure 4.11. The AFF conditions for each of the MgO, MgAl$_2$O$_4$ and Al$_2$O$_3$ crystals are denoted with the marks of +, ◇ and □ respectively. The ψ-AFF conditions can be realized at the positions where these marks coincide at a point or at least very closely. These positions occur at the regions labelled A, B, C, D, ..., whose parameters are listed in table 4.4. At the position A, AFF conditions of the three crystals nearly coincide, which could be the best ψ-AFF condition for taking the images of three crystals at the same time.

4.3.3 Calculated image at ψ-AFF conditions

Using multi-slice many-beam dynamical theory, the intensity distributions at the bottom surface of the MgO, MgAl$_2$O$_4$ and Al$_2$O$_3$ crystals of 50 nm in thickness

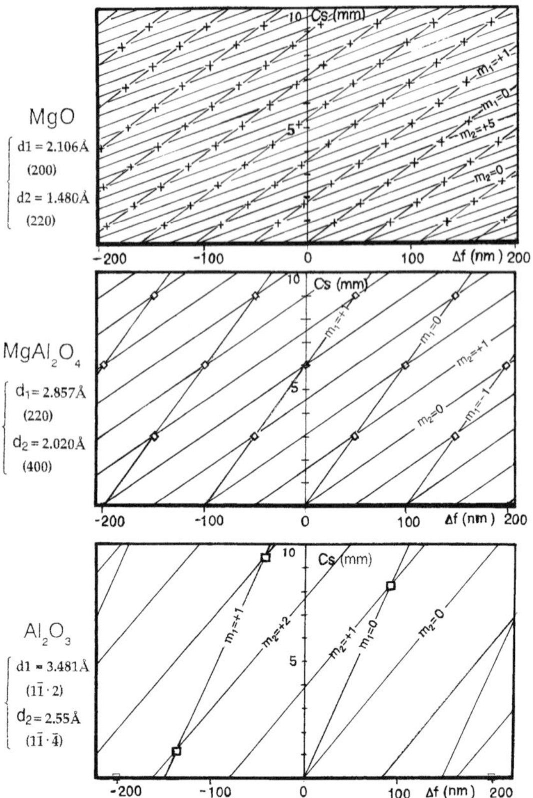

Figure 4.10 C_s–Δf diagram for the AFF condition for (top) {200}, {220} for MgO in the [001] projection, (middle) {220}, {400} for MgAl$_2$O$_4$ in the [001] projection, (bottom) {11.2}, {11.4} for Al$_2$O$_3$ in the [11.0] projection. (After figures 1, 4 and 5 in [14], courtesy Elsevier Science.)

were calculated and compared with the images formed at the ψ-AFF condition. Though the partial coherence of the imaging electrons was taken into account in the present calculations the effect will not be shown in the following figures because it can be deduced from figures 4.6 and 4.7, and it is changed by the system of illumination, for example, by using field emission guns. Figure 4.12 shows the projections of the constituent atom positions of the MgO, MgAl$_2$O$_4$ and Al$_2$O$_3$ crystals in which some atoms overlap. The thin lines in each figure indicate the unit cell size.

As can be seen in figure 4.5, the diffracted waves from the different crystals have different scattering angles. The optimum aperture radius for accepting 200 and 220 type reflections from an MgO crystal is $\phi_1 = 7$ nm^{-1}, which

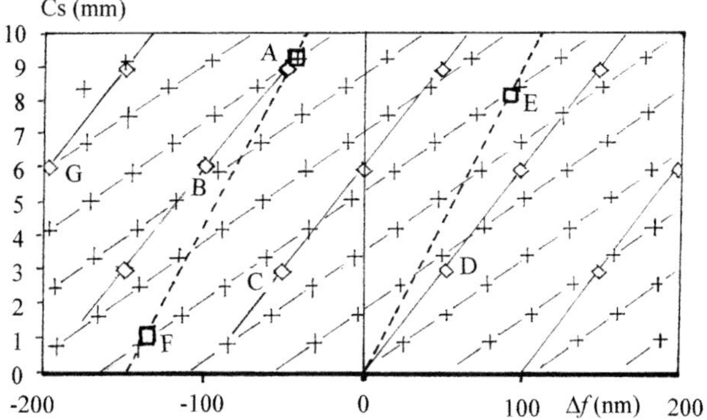

Figure 4.11　Superposition of figures 4.10 (top, middle and bottom). Relationships between the AFF conditions for MgO (+), MgAl$_2$O$_4$ (◇) and Al$_2$O$_3$ (□) are seen. ψ-AFF conditions are realized at the positions A–G. (From figure 7 in [14], courtesy Elsevier Science.)

Table 4.2　AFF condition for MgAl$_2$O$_4$ ($d_1 =$ 0.2857 nm) (220) ($d_2 = 0.2020$ nm) (400).

m_1	m_2	C_s (mm)	Δf (nm)
0	1	3.003 03	49.646
1	3	3.000 2	−49.837
1	4	6.003 2	−0.187
2	6	6.000 3	−99.668
2	7	9.003 4	−50.022
1	5	9.006 25	49.459
0	2	6.006 06	99.293

Table 4.3　AFF condition for Al$_2$O$_3$ ($d_1 =$ 0.3481 nm) (11.2) ($d_2 = 0.2552$ nm) (11.4).

m_1	m_2	C_s (mm)	Δf (nm)
15	28	0.754 78	−2202.976
8	15	0.953 04	−1167.882
1	2	1.151 32	−134.588
1	3	9.408 80	−42.514
−6	−11	1.349 60	899.605
−13	−24	1.547 88	1933.799
16	30	1.906 09	−2337.565

is too large for MgAl$_2$O$_4$ to exclude the reflection 440 and to include only 400 and 220 type reflections. However, as was shown already and will be

Table 4.4 ψ-AFF condition for two or three crystals, MgO, $MgAl_2O_4$, Al_2O_3.

		(m_1, m_2)	C_s (mm)	Δf (nm)
A	Al_2O_3	(1, 3)	9.409	−42.514
	MgO	(6, 23)	9.262	−41.609
	$MgAl_2O_4$	(2, 7)	9.003	−50.022
B	$MgAl_2O_4$	(2, 6)	6.0	−99.6
	MgO	(5, 17)	5.9	−91.0
C	$MgAl_2O_4$	(1, 3)	3.0	−49.8
	MgO	(3, 10)	3.4	−59.8
D	$MgAl_2O_4$	(0, 1)	3.0	49.6
	MgO	(1, 6)	3.4	49.4
E	Al_2O_3	(0, 1)	8.3	92.0
	MgO	(3, 16)	8.4	96.2
F	Al_2O_3	(1, 2)	1.1	−134.5
	MgO	(3, 7)	0.8	−137.8
G	$MgAl_2O_4$	(3, 8)	5.9	−199.1
	MgO	(7, 21)	5.9	−200.0

discussed in the next section, $MgAl_2O_4$ crystals belong to the cubic system and higher order reflections such as 440 can also be in the AFF condition when 220 reflections are in this condition. Thus, when the reflections from 220 and 400 are in the AFF condition with the values $C_s = 9.003$ mm, $\Delta f = -50.02$ nm for $m_1(220) = 2$ and $m_2(400) = 7$, as can be seen in figure 4.10 (middle) and table 4.2, the reflections from {440} also conform to this AFF condition with the value $m_3(440) = 26$. Though the AFF condition ($C_s = 9.003$ mm, $\Delta f = -50.02$ nm) for $MgAl_2O_4$ is slightly different from the AFF condition ($C_s = 9.262$ mm, $\Delta f = -41.61$ nm) for MgO, the image contrast calculated for this condition shown in figure 4.13 indicates that the image contrasts are almost exactly the same as the intensity distribution at the bottom surfaces not only for the $MgAl_2O_4$ crystal but also for the MgO crystal. Mg, O and Al atom positions appear as bright spots which can be seen by comparing figure 4.13 with figure 4.12. Thus the condition $C_s = 9.003$ mm, $\Delta f = -50.02$ nm, $\phi_1 = 7$ nm^{-1} is a good ψ-AFF condition for MgO and $MgAl_2O_4$.

 For images of an Al_2O_3 crystal, an AFF condition is $C_s = 9.409$ mm and $\Delta f = -42.514$ nm for the planes $m_1(11.2)$, $m_2(11.4)$ and attainable only by using the aperture $\phi_4 = 4$ nm^{-1} as can be seen in figure 4.5(c). In this condition, for an $MgAl_2O_4$ crystal, only the diffracted waves from {220} contribute to the image contrast and for an MgO crystal no diffracted waves contribute to the image contrast, as can be seen in figure 4.5(a). The calculated image contrast using $\phi_4 = 4$ nm^{-1} for $MgAl_2O_4$ and Al_2O_3 crystals is shown in figure 4.14, together with the intensity distribution at the bottom surfaces of these

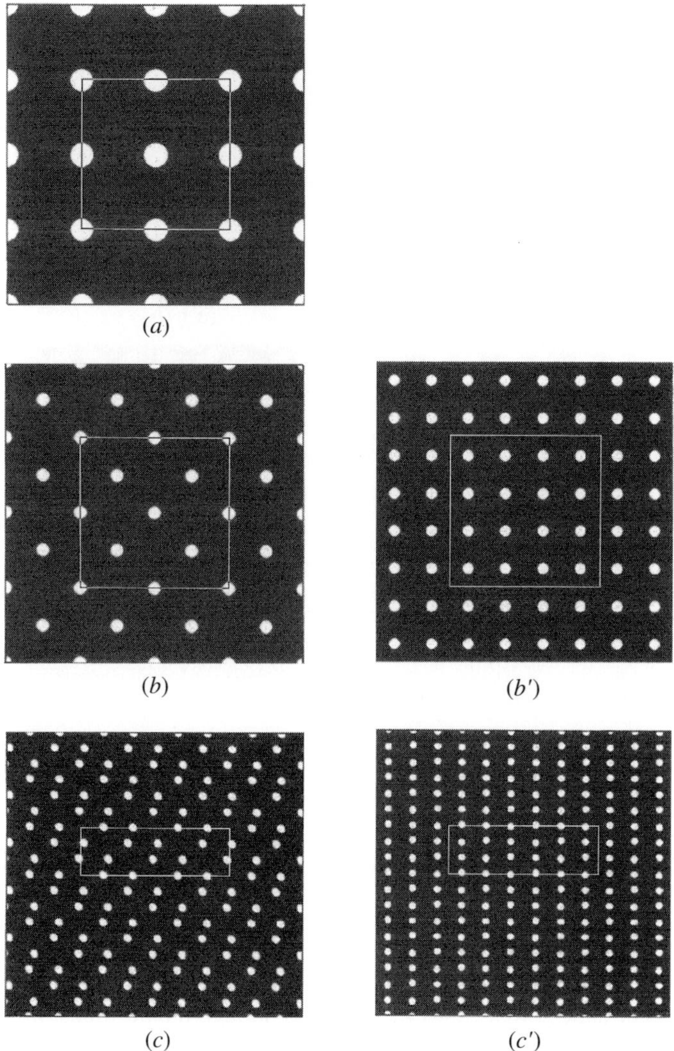

Figure 4.12 (a), (b) (b') and (c) (c') are the projections of the positions of constituent atoms in MgO, MgAl$_2$O$_4$ and Al$_2$O$_3$ crystals onto the a–b, a–b and a–c planes, respectively. Mg atoms are shown in (b), overlapping Al and O atoms in (b'), Al atoms in (c) and O atoms in (c').

crystals. Both image intensities agree well with those at the bottom surfaces, and intensity maxima appear at the atomic positions for the Al$_2$O$_3$ crystal, even though the resolution is not very high. Thus the condition $C_s = 9.404$ mm,

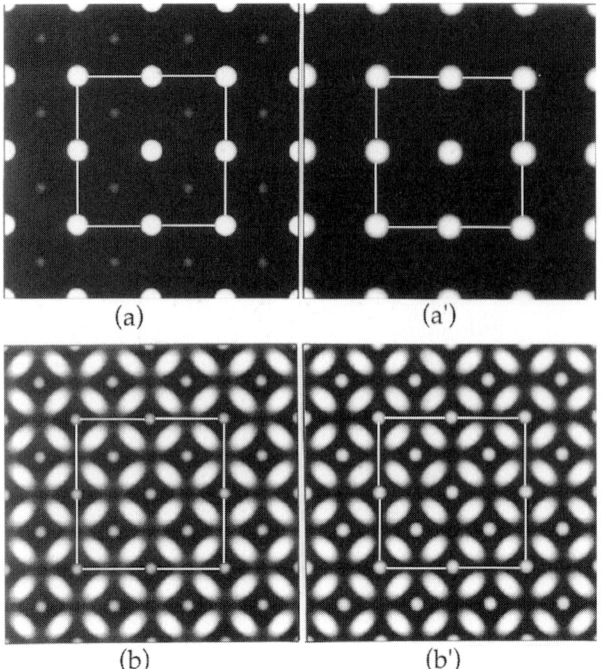

Figure 4.13 (a) and (b) are the calculated image intensities of MgO and $MgAl_2O_4$ for $C_s = 9.003$ mm, $\Delta f = -50.22$ nm and $\phi_1 = 7$ nm^{-1}, which correspond to the exact AFF condition for $MgAl_2O_4$. (a') and (b') are the intensities at the bottom surfaces. 400 kV, thickness $t = 50$ nm, $\Delta = q = 0$. (AFF condition for MgO is $C_s = 9.262$ mm, $\Delta f = -41.609$ nm.)

$\Delta f = -42.514$ nm, $\phi_4 = 4$ nm^{-1} is a good ψ-AFF condition for $MgAl_2O_4$ and Al_2O_3. However, for an $MgAl_2O_4$ crystal, the positions of the Al and O atoms appear dark in figures 4.14(a) and (a'), though the positions of Mg atoms appear bright. This is due to the absence of the 400 type reflections from the images.

For obtaining images from three crystals, MgO, $MgAl_2O_4$ and Al_2O_3, at the same time, an aperture $\phi_3 = 5$ nm^{-1} and the condition $C_s = 9.0034$ mm and $\Delta f = -50.02$ nm were assumed for the calculations of the image contrast and the intensity at the bottom surface; the results are shown in figure 4.15. The positions of the constituent atoms are well displayed for the three crystals, though the fine structure of the calculated image contrast and the intensity at the bottom for Al_2O_3 crystals are slightly different as shown in (c) and (c'). Thus the condition $C_s = 9.0034$ mm and $\Delta f = -50.02$ nm, $\phi_3 = 5$ nm^{-1} is a good ψ-AFF condition for the three crystals. Figure 4.16 shows patterns

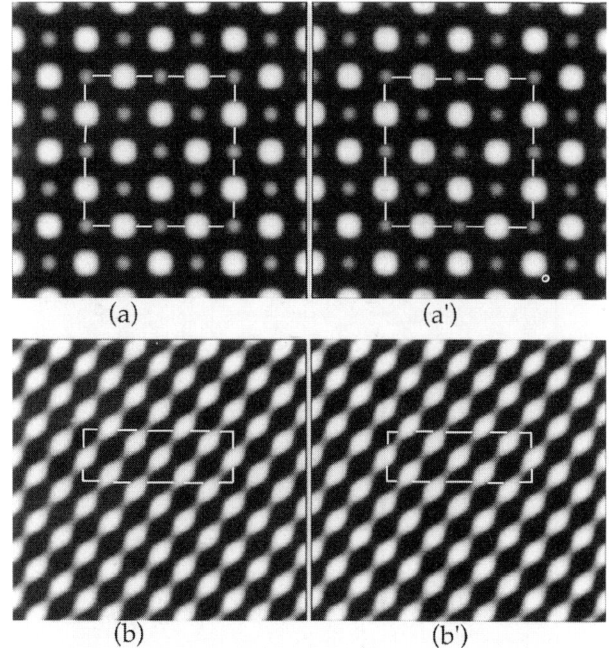

Figure 4.14 (*a*) and (*b*) are the calculated image intensities of $MgAl_2O_4$ and Al_2O_3 for $C_s = 9.409$ mm, $\Delta f = -42.514$ nm and $\phi_4 = 4$ nm^{-1}, which correspond to the exact AFF condition for Al_2O_3. (*a'*) and (*b'*) are the intensities at the bottom surfaces. 400 kV, $t = 50$ nm, $\Delta = q = 0$.

similar to those shown in figure 4.15, but for the condition $C_s = 9.409$ mm, $\Delta f = -42.514$ nm (AFF condition for $m_1(11.2) = 1$, $m_2(11.4) = 3$ but including the 00.6 reflections using $\phi_3 = 5$ nm^{-1}). By comparing the image contrast (*a*), (*b*), (*c*) and the intensity at the bottom surface (*a'*), (*b'*), (*c'*), some similarity exists, but there are some significant differences. This is due to the large aperture $\phi_3 = 5$ nm^{-1}, which allows the diffraction spots from {00.6} to contribute to the image, and thus the AFF condition does not hold for $\phi_3 = 5$ nm^{-1} while it does hold for $\phi_4 = 4$ nm^{-1} for the Al_2O_3 crystal. For the images of the $MgAl_2O_4$ crystal, the Al and O atom positions appear at the correct positions more clearly than in figure 4.14(*a*) which is due to the large aperture, but Mg atoms appear dark. Thus this condition is not an AFF condition for the $MgAl_2O_4$ crystal. The above calculations suggest that the imaging condition of $C_s = 9.0034$ mm, $\Delta f = -50.02$ nm and $\phi_5 = 5$ nm^{-1} is a ψ-AFF condition for taking images of MgO, $MgAl_2O_4$ and Al_2O_3 crystals at the same time, but the condition $C_s = 9.409$ mm, $\Delta f = -42.514$ nm does not hold for $\phi_3 = 5$ nm^{-1} but does hold for $\phi_4 = 4$ nm^{-1}. It should be emphasized that,

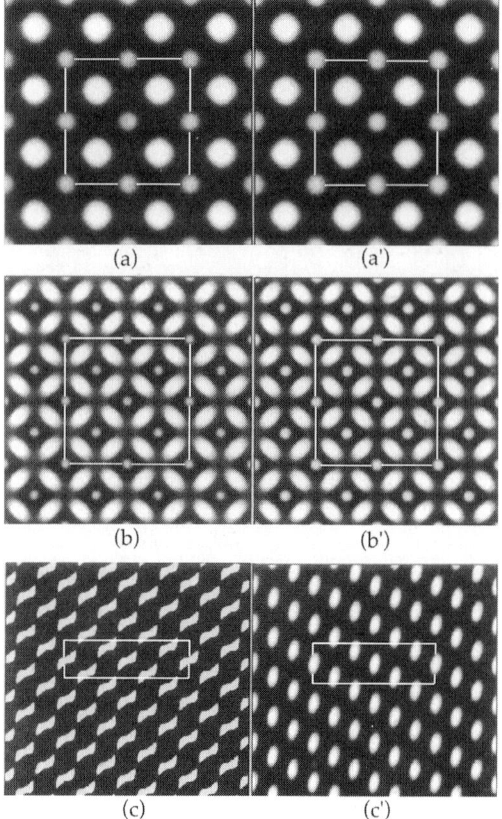

Figure 4.15 Calculated images for (*a*) MgO, (*b*) MgAl$_2$O$_4$, (*c*) Al$_2$O$_3$ for $C_s = 9.0034$ mm, $\Delta f = -50.02$ nm and $\phi_3 = 5$ nm^{-1}, which correspond to the exact AFF condition for MgAl$_2$O$_4$. 400 kV, $t = 50$ nm. (*a'*), (*b'*) and (*c'*) are the intensities at the bottom surfaces.

for obtaining the ψ-AFF condition for more than two crystals, it is important to find appropriate values not only of AFF conditions for individual crystals, but also of the aperture size for selecting diffraction spots suitable for each crystal.

4.4   AFF IMAGES FORMED BY MORE THAN THREE KINDS OF WAVE

As shown in section 4.3.1 in a simple crystal structure, the AFF condition holds for many diffracted waves. An example of this is illustrated by the C_s–Δf

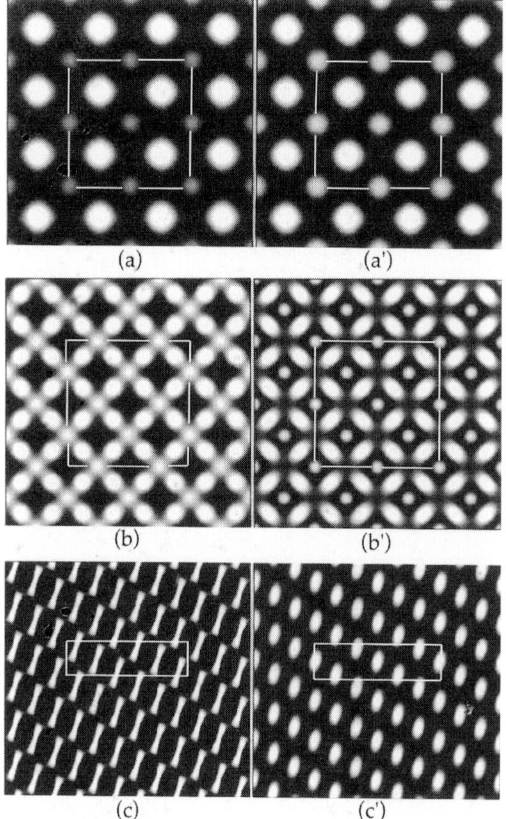

Figure 4.16 Calculated images for (*a*) MgO, (*b*) MgAl$_2$O$_4$, (*c*) Al$_2$O$_3$ for $C_s = 9.409$ mm, $\Delta f = -42.514$ nm and $\phi_3 = 5$ nm^{-1}, which do not correspond to the AFF conditions for Al$_2$O$_3$ and MgAl$_2$O$_4$ because of the inclusion in the aperture of 00.6 and 040 reflections, respectively; (*a'*), (*b'*) and (*c'*) are the intensities at the bottom surfaces.

diagram for an Au crystal with the [110] orientation; the AFF condition is given by $C_s = 0.545$ mm, $\Delta f = 157.3$ nm and is shown for 111, 200, 220 and 311 in figure 4.17. The AFF lines for each reflection are crossing at one point of $C_s = 0.5458$ mm, $\Delta f = 157.3$ nm, for 100 kV electrons. In this condition, 17 diffracted waves have the same phase and contribute correctly to the image. The calculated image contrast for this condition is shown in figure 6 of [5].

As discussed in section 4.3.3 for an Al$_2$O$_3$ crystal in the 11.0 projection, $C_s = 9.409$ mm and $\Delta f = -42.514$ nm correspond to the AFF condition for

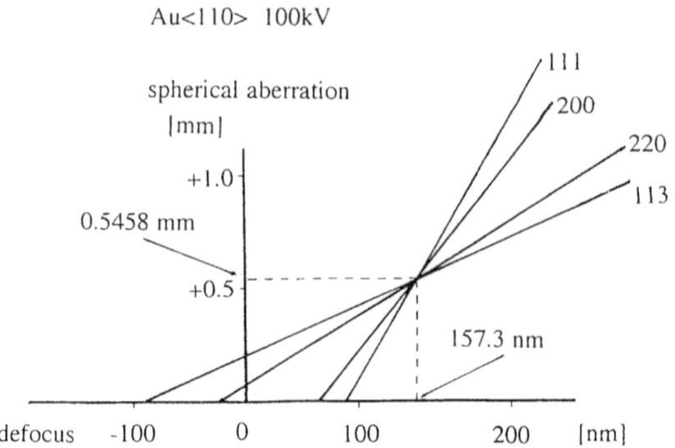

Figure 4.17 C_s–Δf diagram for Au in the [110] orientation at 100 kV, $m_{111} = -3$, $m_{200} = -3$, $m_{220} = 2$, $m_{222} = 15$, $m_{113} = 11$, $C_s = 0.5458$ mm, $\Delta f = 157.3$ nm. (After figure 13 in [14], courtesy Elsevier Science.)

{11.2}, {11.4}, but not for {00.6}. However, another AFF condition for {11.2}, {11.4}, with $C_s = 1.15$ mm, $\Delta f = -134.59$ nm, can be an AFF condition for {00.6} and {22.4} also. Thus as illustrated in figure 4.7, even though the large aperture $\phi_2 = 6$ nm^{-1} was used, the AFF condition for Al_2O_3 holds.

The values C_s and Δf for the AFF condition for more than three lattice planes can be obtained by the use of algebra [11]. However, as shown in the present paper, it is easier to select many pairs of lattice planes with the same values of C_s and Δf for AFF conditions by solving the pairs of equations (4.6) for many reflections with systematic changes of the values m.

4.5 MECHANISM FOR CHANGING THE SPHERICAL ABERRATION COEFFICIENT

Adjustment of C_s values can be carried out by changing the specimen height, Z, from the top surface of the bottom part of the objective electron lens. Figures 4.18(*a*) and (*b*) show the variation of C_s against Z for the JEM-4000EX at Osaka University and the JEM-1000ARM at UC—Berkeley for various voltages from 200 to 1000 kV, respectively. Since most of the conventional microscopes have the mechanism available for changing the specimen position along the optical axis, it is not difficult to change the value Z with an accuracy of about 10 μm, using for example a micro-meter. A simple means of changing the specimen position may be by the use of spacers with different thicknesses when the specimen is mounted on the cartridge.

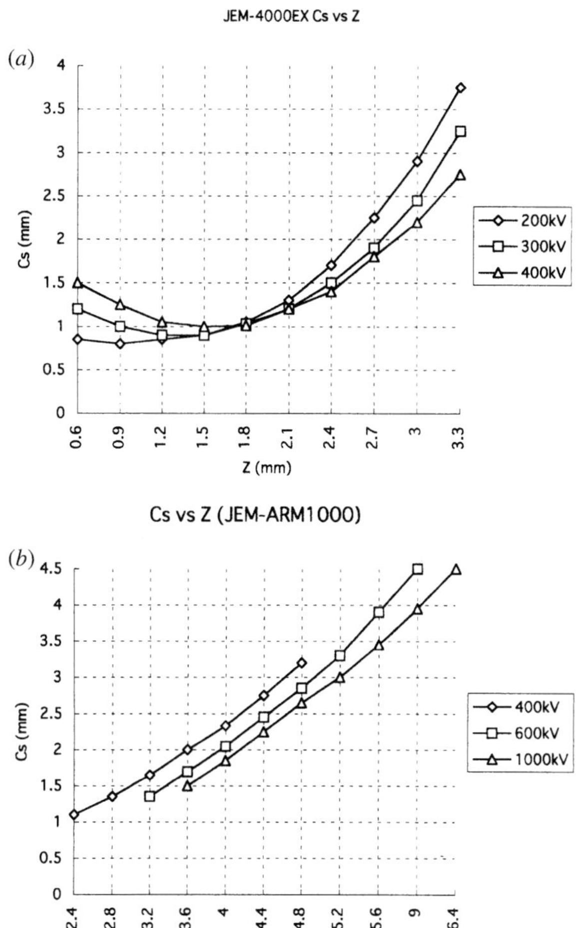

Figure 4.18 Variation of C_s against Z for (*a*) JEM-4000EX, (*b*) JEM-1000ARM at voltages of 200–1000 kV.

4.6 COMMENTS AND ACKNOWLEDGMENTS

In order to study the atomic arrangement at interfaces, which are generally subject to complicated atomic displacements, a proper understanding of the atomic structure of the crystals on both sides of an interface is essential. This will be of help in the simulation of the structure of an interface [12].

As can be seen in figure 4.10, AFF conditions are realized for $C_s = 0$, $\Delta f = 0$, and thus electron microscopes with very small values of C_s are very useful for taking high resolution images of general crystals with several tens of nanometres in thickness at very small values of defocus. It is not necessary to use very high voltages to make the wavelength of the illuminating electrons very small.

The approach to make C_s small by use of a quadrupole lens [13] may be a useful means for attaining an effective ψ-AFF condition.

The preparation of the present article was helped greatly by the efforts of Professors H Endoh and M Hashimoto to whom the present author would like to express his sincere thanks. Some parts in this chapter are taken from a paper in *Micron* [14], courtesy Elsevier Science.

REFERENCES

[1]   Kuwabara M, Endoh H, Tubokawa Y, Hashimoto H, Yokota Y and Simizu R 1985 *Proc. Int. Symp. on Behaviour of Lattice Imperfections in Materials—In Situ Experiments with High Voltage Electron Microscopes (Osaka, 1985)* ed H Fujita (Osaka: Osaka University, Research Centre for Ultra-High Voltage Electron Microscopy) pp 341–50

[2]   Hashimoto H, Endoh H, Tomita M, Ajika N, Kuwabara M, Hata Y, Tubokawa Y, Honda T, Harada Y, Sakurai S, Etoh T and Yokota Y 1986 *J. Electron Microsc. Tech.* **3** 5–24

[3]   Scherzer O 1949 *J. Appl. Phys.* **20** 20

[4]   Ishizuka K 1980 *Ultramicroscopy* **5** 55

[5]   Hashimoto H, Endoh H, Takai Y, Tomita H and Yokota Y 1978/1979 *Chem. Scr.* **14** 23–31

[6]   Hashimoto H, Makita Y and Endoh H 1996 *Mater. Chem. Phys.* **46** 7

[7]   Cowley J M and Moodie A F 1957 *Proc. Phys. Soc.* **70** 505

[8]   Yamaji K, Yokota Y and Hashimoto H 1990 Private communication

[9]   Sukedai E, Liu W, Mabuchi H, Hashimoto H and Nakayama Y 1992 *Proc. 5th Asia Pacific Conf. on Electron Microscopy (Beijing, 1992)* ed K H Kuo and Z H Zhai (Singapore: World Scientific) pp 316–7

[10] Hesse H, Senz S, Scholz R, Werner P and Heydenreich J 1994 *Interface Sci.* **2** 221–73

[11] Endoh H 1984 *PhD Thesis* Osaka University

[12] Hutchison J L, Chou C T, Casanove M J, Cherns D, Steeds J W, Ashenford D A and Lunn B 1994 *Ultramicroscopy* **53** 91

[13] Krivanek O L, Dellby N, Spence A J, Camps R A and Brown L M 1997 *EMAG'97 (Cambridge, 1997) (Inst. Phys. Conf. Ser. 153)* (Bristol: Institute of Physics) p 35

[14] Hashimoto H, Endoh H, Hashimoto M, Luo Z P and Song M F 1998 *Micron* **29** 113–21

5

PROBING ATOMIC BONDING USING FAST ELECTRONS

Colin J Humphreys and Gianluigi A Botton

5.1   INTRODUCTION

Many of the properties of materials are determined by their electronic structure, in particular by the structure of the outer electron orbitals which can be described by bonds in real space and by energy bands in reciprocal space. In order to understand existing materials, and to tailor-make new and improved materials for specific applications, it is therefore important to determine their electronic structure. Electron diffraction and electron energy loss spectroscopy (EELS) are key techniques for probing the bonding in solids and M J Whelan has made immense contributions to these fields, from formulating the Howie–Whelan equations of the dynamical electron diffraction theory [1] to pioneering work on the electronic structure of various forms of carbon using EELS [2].

Understanding the electronic structure of solids is not only important for explaining electrical and optical properties of materials (for example why metallic copper is a better electrical conductor than metallic aluminium, or why dislocations quench light emission in GaAs but not in GaN); it is also important for explaining structural properties of materials (for example why copper is ductile and silicon is brittle, or why bismuth embrittles grain boundaries in copper and silver does not).

In this chapter we will review the field of electronic structure determination in solids with particular emphasis on recent developments.

5.2   THE CRYSTAL POTENTIAL IN ELECTRON DIFFRACTION

An incident electron beam undergoes elastic and inelastic scattering in a crystal. These effects are represented by an 'optical' potential, which is the potential

65

that the incident electrons effectively see, of the form

$$V^{\mathrm{opt}}(r) = V(r) + iV^i(r) + \Delta V(r) \tag{5.1}$$

where $V(r)$ is due to elastic scattering, $V^i(r)$ is due to inelastic scattering and $\Delta V(r)$ is a real part addition to the potential due to virtual inelastic scattering. It can be shown that $\Delta V(r)$ is negligible for incident electron energies greater than 100 keV. The elastic part of the potential, $V(r)$, is due to both Coulomb interaction and exchange; however, it can be shown that the exchange terms are negligible for incident electron energies over 50 keV. Thus we have the important result that for incident electron energies above 100 keV exchange (between the fast electron and crystal electrons) and virtual inelastic scattering are negligible and the interaction of the fast electron with the crystal is simply coulombic [3]. The fast electron therefore 'sees' the crystal potential $V(r)$ including all solid state bonding effects and exchange and correlation effects among the crystal electrons, hence the importance of fast electrons as a probe for measuring solid state bonding effects.

In a perfect crystal we can expand the crystal potential as a Fourier series based on the reciprocal lattice:

$$V(r) = \sum_g V_g \exp(2\pi igr). \tag{5.2}$$

For a centrosymmetric crystal the V_g are real and for a non-centro crystal the V_g are complex. The problem in determining the bonding is to measure accurately the magnitude and phase of V_g for a sufficient number of reflections g that equation (5.2) converges. V_g is related to the structure amplitude F_g by

$$V_g = \frac{h^2}{2\pi me\Omega} F_g \tag{5.3}$$

where Ω is the unit cell volume.

5.3   THE PHASE PROBLEM

We can write

$$F_g = |F_g| \exp(i\varphi_g) \tag{5.4}$$

where φ_g is the phase of F_g with respect to a given origin. The measurement of φ_g is a well known problem which arises because on kinematical or two-beam dynamical theory for electron or x-ray diffraction:

$$I_g \propto |F_g|^2 \tag{5.5}$$

where I_g is the intensity of the gth diffracted beam (in the case of x-rays, F_g is the x-ray structure factor instead of the electron structure amplitude). Hence the diffracted intensity I_g contains no information on the phase of F_g.

In many-beam dynamical theory, however, the phase information is not lost since the intensities are a function of the phases. For example, in three-beam dynamical theory, for beams 0, g and h, the intensity of the gth diffracted beam is

$$I_g = \mathrm{fn}(F_g, F_h, F_{h-g}, \varphi_g, \varphi_h, \varphi_{h-g}). \tag{5.6}$$

It has been known for some time that analytical solutions can be written down for the diffracted beam intensities I_0, I_g and I_h in three-beam dynamical theory but the solutions are too complex to extract the phases. Hence although the intensities of many-beam electron diffraction patterns contain phase information, the problem lies in extracting this information.

However, two-beam dynamical theory equations, for example the Howie–Whelan equations [1], have a very simple analytical form. This suggests that if we can find conditions which reduce the three-beam dynamical theory equations (which contain phase information) to the much simpler two-beam form then we may be able to extract the phase information. There are three ways in principle in which this can be done, outlined below.

5.3.1 The critical voltage effect

In its original form, the critical voltage effect involved exciting a systematic row of reflections and orienting the crystal to a second order Bragg reflecting condition, so that the reflections 0, g and $2g$ were mainly excited. At a particular incident electron accelerating voltage, the critical voltage, it is found that the intensity I_{2g} is a minimum [4]. Lally *et al* [5] showed that at the critical voltage, two eigenvalues of the fast electron in the crystal are degenerate, i.e.

$$\gamma^{(2)} = \gamma^{(3)}. \tag{5.7}$$

In addition Bloch waves 2 and 3 destructively interfere. Hence at the critical voltage, three Bloch waves essentially reduce to two, and the three-beam equations reduce to two-beam form. This method is excellent for measuring accurately the magnitudes of low order structure factors (see later) but it is very restrictive since only low order structure factors can be determined and the method only works for centrosymmetric crystals.

5.3.2 The Gjønnes–Høier point

Gjønnes and Høier [6] modified the critical voltage effect, and applied their theory to the general three-beam case where the reflections 0, g and h are not collinear. They showed that there was one, and only one, angle of incidence for which $\gamma^{(2)} = \gamma^{(3)}$, and that by measuring the splitting of intersecting Kikuchi lines the magnitude of scattering factors could be measured. However, the splitting is small and the method is not very accurate, particularly for phases. In this method, three-beam theory reduces to two-beam form at a single point in reciprocal space, the Gjønnes–Høier point corresponding to the particular orientation for which $\gamma^{(2)} = \gamma^{(3)}$.

5.3.3   Inversion of three-beam convergent beam electron diffraction (CBED) patterns

Recently a new method has been developed for measuring phases by inverting three-beam convergent beam electron diffraction (CBED) patterns [7]. The method again depends on reducing the complex three-beam equations to a two-beam form, but instead of using the condition that two eigenvalues are degenerate, which leads to very restrictive cases, we use the condition that the third Bloch wave is not excited so that three-beam theory necessarily reduces to two-beam form.

In particular, in the Bloch wave formalism we can write the fast electron wavefunction in the crystal as

$$\psi(r) = \sum_{j} \alpha^{(j)} \sum_{g} C_g^{(j)} \exp\left\{2\pi i\left(k^{(j)} + g\right)r\right\} \tag{5.8}$$

using the conventional notation where $\alpha^{(j)}$ is the excitation amplitude of Bloch wave j and $C_g^{(j)}$ are the eigenvectors [8]. The application of standard boundary conditions gives

$$\alpha^{(j)} = C_0^{(j)\,*}. \tag{5.9}$$

In the three-beam case, it is evident that if $C_0^{(3)\,*} = 0$ then wave 3 is not excited for all beams 0, g and h. Moodie *et al* [7] have shown that setting $C_0^{(3)\,*} = 0$ in equation (5.8) defines (in the three-beam case) a line on each CBED disc 0, g and h along which the intensities I_0, I_g and I_h have two-beam form.

Setting $C_g^{(3)} = 0$ defines a different line on CBED disc g along which I_g has two-beam form. Setting $C_h^{(3)} = 0$ defines a different line on disc h along which I_h has two-beam form. Thus on a three-beam CBED pattern, five lines can be drawn along which the discs have two-beam form. By measuring the separations of these lines (e.g. with a ruler or a computer using digitized CBED patterns), F_g, F_h and F_{g-h} can all be measured and also the phases. This is a direct method which gives unique results. The phases can be particularly accurately determined. By choosing different sets of three-beam patterns from the same material it is possible to 'step out' in reciprocal space and determine phases for a complete set of reflections, out to any desired limit.

Initial experimental results [9] confirm that both the phase-invariant and the magnitudes of the three structure amplitudes can be measured directly, uniquely and simultaneously from the intensity distribution in a single diffraction pattern. The measurements are independent of crystal thickness and the phase and magnitudes are measured independently.

5.4   MEASURING ACCURATE STRUCTURE AMPLITUDES

Preliminary experimental work [9] shows that although the three-beam inversion method described above is very powerful for determining phases, the structure amplitudes are measured to only limited accuracy because of the precision with which the two-beam loci can be identified on the CBED patterns. We therefore need to explore alternative techniques for measuring the magnitude of the structure amplitudes.

X-ray diffraction measurements of charge densities do not have sufficient accuracy to detect solid state bonding effects except in very special cases, for example single crystal silicon which can be grown dislocation free and from which x-ray interferometers can be made. Using such an interferometer, Aldred and Hart [10] determined Si scattering factors to 0.07%.

Hewat and Humphreys [11] showed that the critical voltage effect could be used to measure low index scattering amplitudes (for electrons) in Si, which when converted to (x-ray) scattering factors had an accuracy of 0.1% and agreed with the Aldred–Hart values to within this experimental error, thus establishing electron diffraction methods for measuring bonding charge densities. The critical voltage effect, being a null method, is an excellent technique to measure structure amplitudes, but it is limited to low index structure amplitudes and the main source of error is the uncertainty in the Debye–Waller factor, which is usually not known very accurately. Smart and Humphreys [12] combined low index structure amplitudes measured by the critical voltage effect with higher index structure factors measured by x-ray diffraction to plot bonding charge densities for a range of materials including Si, Ge, Al, Cu, Fe and Ni.

The development of commercial energy filters enabled energy-filtered CBED patterns to be formed, which could then be matched quantitatively to theoretical simulations. (Although the effect of the inelastic scattering on the elastic scattering of electrons by crystals can phenomenologically be taken into account by adding an imaginary part to the potential, this does not account for the contribution that inelastically scattered electrons make to the image when they pass through the objective aperture. This contribution is difficult to calculate accurately, hence the importance of removing it using an energy filter.) Quantitatively matching experimental and theoretical CBED patterns involves regarding the Fourier coefficients of crystal potential as variables and the use of computer fitting procedures. This method was pioneered by Spence and Zuo [13]. Saunders *et al* [14] generalized the systematic row refinement method of Spence and Zuo [13] to using zone-axis CBED patterns and matching these quantitatively.

Limitations of the CBED matching procedure include possible uniqueness problems (is the set of V_g values determined the correct set?) due to false minima in computer fitting routines, and the need to know accurately Debye–Waller factors. However, the CBED method is very powerful.

5.5 EELS AND DENSITY FUNCTIONAL THEORY

Recently a powerful new method has been developed to determine the bonding in materials based on a combination of experimental EELS and density functional theory (DFT) [15]. This work builds on the pioneering work of Egerton and Whelan [2]. The new technique is not sensitive to Debye–Waller factors or to false minima in computer fits. We will describe the use of this technique in the problem of determining the bonding in the intermetallics FeAl, CoAl and NiAl, which all have the same B2 crystal structure, but which have very different mechanical properties. Although DFT is required to interpret the EELS spectra in detail, many of the EELS results can be interpreted simply and quickly by inspection, as described later. First we comment on DFT methods.

5.6 DENSITY FUNCTIONAL THEORY

The development of the DFT which is now included in various band structure techniques (linear muffin tin orbitals (LMTO), linearized augmented plane waves (LAPW), pseudo-potentials etc) has made these methods faster and more reliable, and the power of workstations has increased 100-fold in the last six years, so that these computer codes can now be executed on workstations and even on fast personal computers. We have used the LMTO method with the local density approximation (LDA) for exchange and correlation. The LMTO calculations were run self-consistently and EELS spectra were calculated by combining the computed density of states (DOS) with the matrix elements for the transition from initial to final states (see below). This treatment is based on the single particle approximation of the EELS spectrum (see, for example, [16]). We have also calculated spectra based on the more computationally demanding Korringa–Kohn–Rostoker (KKR) non-linear method for comparison (agreement is very good for the systems studied).

5.7 ELECTRON ENERGY LOSS SPECTROSCOPY (EELS)

In EELS, the fast incident electron excites electrons in the crystal from bound to unoccupied states (that is, states above the Fermi level). If the initial state has angular momentum L then the final states must have angular momentum $L + 1$ or $L - 1$. Hence if the initial state is a p state, the final state must be d or s. The EELS intensity for a transition from an initial state with angular momentum L to a final unoccupied state with angular momentum $L + 1$ is

$$I \propto |M_{L+1}|^2 \rho_L^* \rho_{L+1} \tag{5.10}$$

where ρ_L and ρ_{L+1} are the density of the initial states and the density of unoccupied final states, respectively. For excitations from inner shells the density of initial states, ρ_L, is a delta function to a good approximation, hence

$$I \propto |M_{L+1}|^2 \rho_{L+1}. \tag{5.11}$$

Thus the intensity of EELS spectra is proportional to the density of unoccupied final states. This is an important result.

The matrix element M_{L+1} is given by

$$M_{L+1} = \int \psi_L^*(r) H \psi_{L+1}(r) \, \mathrm{d}r \tag{5.12}$$

where $\psi_L(r)$ and $\psi_{L+1}(r)$ are the initial and final state wavefunctions and H represents the electron–electron interaction. For non-zero M_{L+1} then ψ_L and ψ_{L+1} must overlap, hence if ψ_L is a localized core state, ψ_{L+1} must be local to the same atomic volume, hence EELS probes the local DOS.

For intermetallics based on transition metal (TM) elements (e.g. NiAl) we are particularly interested in the number of 3d electrons, hence in the transitions from initial 2p states to unoccupied 3d states. The 2p state is split into $2p^{3/2}$ and $2p^{1/2}$ initial states (spin–orbit coupled), hence the transition from 2p to 3d gives rise to two distinct sets of peaks, separated in energy (the $2p^{1/2} \rightarrow$ 3d transition gives rise to the L_2 peak and the $2p^{3/2} \rightarrow$ 3d transition to the lower energy L_3 peak). The L_2 and L_3 peaks are known as white lines since they appear on x-ray absorption spectra as white lines on a dark background.

5.8   Ni AND NiAl: THE BONDING COMPARED

Figure 5.1(*a*) shows experimental EELS spectra for the Ni L_{2-3} edges from f.c.c. Ni and from NiAl. It is immediately apparent that EELS is sensitive to the bonding changes which occur when Ni goes from metallic Ni to NiAl. Figure 5.1(*b*) shows the LMTO calculated spectra, which agree well with the experimental spectra (except at higher energies about 10 eV above the L_2 edge where the LMTO method does not consider enough basis functions). The more computer intensive KKR calculations are valid at higher energies and they not only agree with the experimental results and with the LMTO calculations to about 10 eV above the L_2 edge, but also accurately reproduce the experimental intensity up to 50 eV above the L_2 edge [16].

The Hume-Rothery rule for the valence of Ni when it is an 'electron compound', as in NiAl, is that Ni has a full 3d band of ten electrons and an empty 4s band, so that its valence is zero. We can see from figure 5.1(*a*) that this is a simplistic way of considering the bonding in these materials. If Ni in NiAl had a full d band then there would be no L_{2-3} white lines. Hence from figure 5.1 there are unoccupied Ni d states in NiAl.

We can also see from figure 5.1(*a*) that the intensity of the Ni L_2 and L_3 white lines is stronger in metallic Ni than in NiAl. Since the EELS intensity is

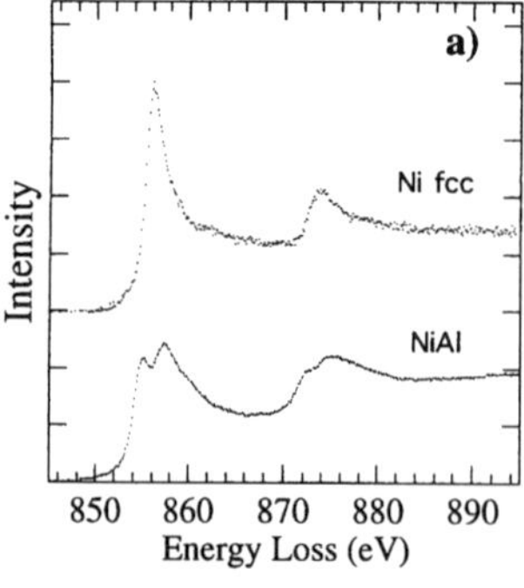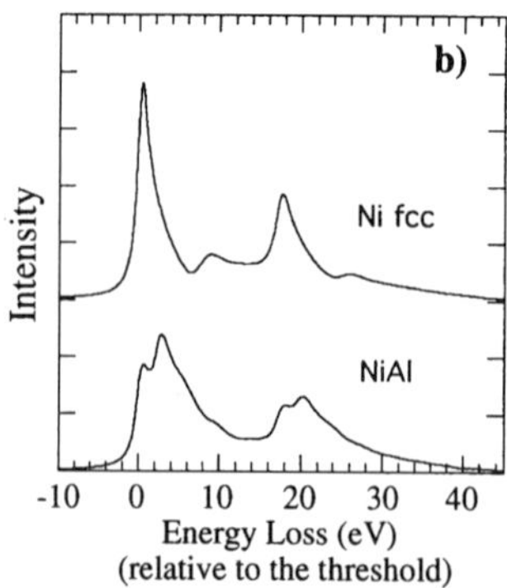

Figure 5.1   Ni L_{2-3} edges in NiAl and f.c.c. Ni. Experimental EELS spectra (*a*) and theoretical calculations using the LMTO technique (*b*).

proportional to the density of unoccupied states it follows that there are more unoccupied Ni 3d states in metallic Ni than in NiAl at the Fermi energy. This effect is due to the fact that Ni 3d electrons hybridize with Al electrons when Ni is alloyed to become NiAl. Such an effect is confirmed by the analysis of the Al L_{2-3} edge which shows that Al s states hybridize with the Ni d bands and this results in the loss of 'free metallic' character near the Fermi energy. At the same time, the experimental spectra and the calculations of the Al L_{2-3} edge show that there are common states between the Al and Ni atoms and this suggests that there is covalency in the bonds of NiAl [15].

5.9   NiAl, CoAl AND FeAl: THE BONDING COMPARED

The TM aluminides NiAl, CoAl and FeAl have the same crystal structure (B2) but very different mechanical properties. It is therefore of interest to see whether EELS can detect changes in bonding in this series of neighbouring element TM aluminides. Since these materials have the same crystal structure they constitute an ideal test case to study systematic changes in the electronic structure which can potentially be related to changes in the mechanical or magnetic properties.

Figure 5.2 shows the experimental EELS L_2 and L_3 edges for Fe in FeAl, Co in CoAl and Ni in NiAl. The spectra have been recorded from regions of the same specimen thickness relative to the total inelastic mean free path so that any multiple inelastic scattering effects are comparable in the materials. Again, DFT LMTO calculations match well these experimental spectra [15] particularly for CoAl and NiAl. It is evident from figure 5.2 that the intensity of the Co L_{2-3} lines is greater than that of the Ni L_{2-3} lines. Since the intensity of these EELS spectra is proportional to the density of unoccupied 3d states

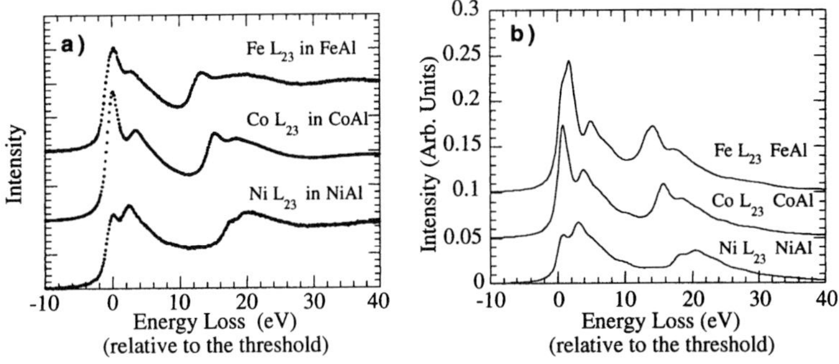

Figure 5.2   TM L$_{2-3}$ edges in FeAl, CoAl and NiAl. Experimental edges obtained with EELS (*a*) and theoretical edges calculated with the LMTO method (*b*).

we can immediately deduce that the TM d band fills on going from CoAl to NiAl. For FeAl, the single particle calculations using equation (5.10) and shown in figure 5.2(*b*) suggest a similar trend, that the TM d band fills on going from FeAl to CoAl. In the experiments, however, it is clear that this ground state effect is masked in part by the fact that excitation effects start to become more important for FeAl (see section 5.12). This is shown by the fact that the agreement between theory and experiment is not as good as in NiAl (e.g. the position of the shoulder to the white line moves to lower energy instead of higher energy) and there is also a broadening of the white line relative to the Co L$_{2-3}$ edge. Preliminary calculations accounting for some excitation effects (in particular a core hole) show that the intensity at the threshold is lower than expected from a simple extrapolation of the trend observed on going from NiAl to CoAl. Hence the Fermi energy increases on going from FeAl to CoAl to NiAl. In fact the Ni d band in NiAl is nearly full, so Hume-Rothery was nearly right!

It is clear from figures 5.1 and 5.2 that the TM L$_2$ and L$_3$ peaks of CoAl and NiAl (and to a lesser extent FeAl) are each split into sub-peaks, whereas the L$_2$ and L$_3$ peaks from metallic Ni are not split. This can be explained from a consideration of the calculated DOS for the TM d band of the TM aluminides, which is a complicated function of energy with a number of oscillations [15]. The 3d DOS for the TMs Ni, Co and Fe in the TM aluminides are broadly similar, but the position of the Fermi energy moves significantly, increasing from FeAl through CoAl to NiAl. Since the intensity of the EELS L$_{2-3}$ edges is proportional to the DOS above the Fermi energy, the split L$_2$ and L$_3$ peaks correspond to oscillations of the DOS above the Fermi energy. The reversal of the relative heights of the split peaks in CoAl and NiAl (figure 5.2) is because the Fermi energy shifts and in CoAl it is near a maximum of an oscillation in the DOS, hence the first peak is high, whereas in NiAl it is near a minimum,

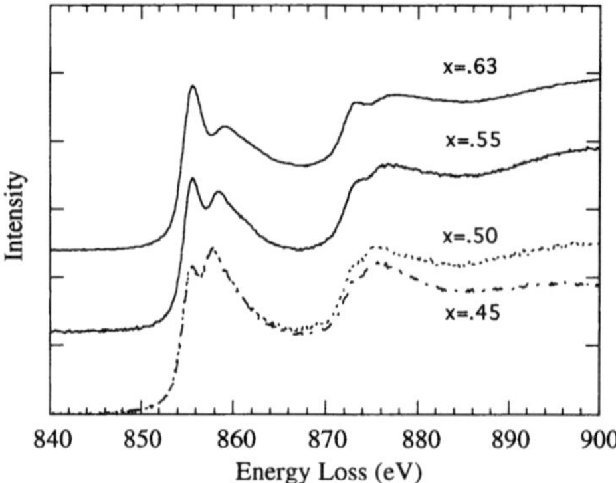

Figure 5.3 EELS Ni L_{2-3} edges in non-stoichiometric Ni_xAl_{1-x} as a function of composition.

hence the first peak is low [15]. For these compounds this peak represents an e_g band and thus orbitals pointing towards Al.

We have already shown that in going from metallic Ni to NiAl electrons in the Ni d band hybridize. Similarly, by comparing EELS spectra from FeAl with those from metallic Fe, and from CoAl with those from metallic Co, it is clear that in each case the d bands of the TM are affected in the same way.

Where do these hybridizing electrons come from? Al K-edge EELS spectra from metallic Al, FeAl, CoAl and NiAl show a pre-peak near the K edge for FeAl, CoAl and NiAl, but not for metallic Al. Calculations confirm that this pre-peak is due to transitions from Al s to unoccupied 2p states in FeAl, CoAl and NiAl and this peak occurs at the same energy (relative to E_F) where the first peak in the TM edges is present. This observation and the results at the Al L_{2-3} edges in these compounds therefore represent the introduction of a directional covalent component to the Al–TM bonds. Hence there is a covalent component to the bond between Ni and Al, Co and Al and Fe and Al and, as the Fermi energy moves in the band where the Al p and TM d bonds occur, the details of the bonding characteristics in these materials will vary systematically.

5.10 NON-STOICHIOMETRIC Ni_xAl_{1-x}

Figure 5.3 shows EELS spectra from non-stoichiometric NiAl. It is clear that on the Ni-rich side the L_{2-3} peak intensity continuously varies with composition and provides a fingerprint of the local composition to about 1% compositional accuracy, with a spatial resolution that of the electron probe size, less than 1 nm in a field emission gun instrument. On the Ni-deficit side, the Ni L_{2-3} intensity is constant, although the background at higher losses varies as shown in figure 5.3.

Why does the Ni L_{2-3} EELS spectrum from non-stoichiometric NiAl vary in this way? On the Ni-rich side, there is evidence that the additional Ni atoms replace Al atoms and sit on the Al sub-lattice. A simple starting point to interpret the variation in the spectra is the following. Since almost zero-valence Ni is replacing trivalent Al, the Fermi energy decreases within the band and this results in it moving up a peak in the Ni DOS curve [17]. The first split peak of both the L_3 and L_2 white lines therefore increases as shown in figure 5.3. A more detailed analysis is given elsewhere [18].

In addition, we have calculated the Ni L_{2-3} EELS spectrum for Ni atoms on the Al sub-lattice. The spectrum has a very strong first sub-peak and an almost absent second sub-peak for both the L_3 and L_2 white lines. By comparison, for Ni on Ni sites in NiAl, the first sub-peak is weaker than the second. Hence EELS can be used to determine site occupancies.

On the Ni-deficit side, Lipson and Taylor pointed out as long ago as 1939 [19] that if Al has its usual valence of three and the valence of Ni is taken to be zero, then as Ni is removed from NiAl, leaving Ni vacancies, the average number of valence electrons per atom increases, and the number of atoms per unit cell decreases, but the number of valence electrons per unit cell remains constant (at 3). This is consistent with our observation of similar spectra in Ni-deficient alloys, and consistent with the conclusions of Cottrell [20] that there is no significant change in the bonding when constitutional vacancies are introduced in the Ni sub-lattice.

5.11   PHASE STABILITY

A high DOS at the Fermi energy usually results in low phase stability because the total electron energy is high [21]. We can fix the position of E_F on the DOS curve from DFT and confirm this using EELS. For the alloys considered in this paper, the DOS at the Fermi energy is highest for FeAl and lowest for NiAl, with CoAl lying between.

Experimentally, all these alloys are ordered up to their melting point, but FeAl exhibits the highest degree of local disorder as is evident from diffraction patterns in which there is significant diffuse scattering from FeAl. The diffuse scattering is due to small atomic displacements, probably induced by thermal vacancies which have a particularly high density in FeAl. Thus EELS can be used to study phase stability.

5.12   TiAl AND THE SINGLE PARTICLE APPROXIMATION

We have shown above the good agreement between theory and experiment for EELS spectra from NiAl and CoAl. The agreement is less good for FeAl and significantly less good for TiAl [16] due to the breakdown of the single particle approximation used in writing down the equations in section 5.3.

The problem arises for two main reasons.

1. When an electron is excited from a 2p state to an empty 3d state it leaves behind a hole in the 2p state. This core hole interacts with the excited electron, but this interaction is neglected in the single particle approximation. The reason this approximation works well for EELS spectra from NiAl and CoAl is because they have many 3d electrons which screen the excited electron from the core hole. However, TiAl has fewer 3d electrons and hence the screening is less effective. This effect starts to be seen in FeAl as discussed above.

2. The single particle approximation neglects the d–d electron interaction between the excited electron and the other d band electrons (although the theory takes into account exchange and correlation effects between all electrons in the ground state wavefunction). In the late TM aluminides (NiAl and CoAl) there are a limited number of unoccupied states available to the excited electron, so exchange and correlation effects with this electron are small. However, in the early TM aluminides (for example, TiAl) due to the relatively empty 3d band exchange and correlation effects are large.

Hence for the quantitative interpretation of EELS spectra from early TM aluminides, the single particle approximation partially breaks down and a theory is required which fully takes into account not only the ground state but also the excited state and the excitation process, including all exchange and correlation effects and the core hole.

5.13   CONCLUSIONS

Electron diffraction and EELS are powerful methods to probe the bonding between atoms in crystals. We have shown that a new three-beam CBED inversion method can measure structure amplitude phases accurately, directly and uniquely. The critical voltage technique and energy filtered CBED patterns plus simulations can be used to measure accurately the magnitudes of structure amplitudes. EELS and DFT provide a powerful combination to characterize in detail the bonding in solids. A particularly simple interpretation of EELS spectra is possible if the single particle approximation holds, and we have demonstrated that it does hold for NiAl, CoAl and FeAl. In these three aluminides there are common states and strong hybridization between the Al and TM atoms giving rise to a covalent component of the bonding between Al and the TM atoms. Deviations from stoichiometry in NiAl are revealed by EELS, and the spectra confirm that on the Ni-rich side the extra Ni atoms sit on Al sites whereas on the Ni-deficit side there are Ni vacancies. Information concerning the phase stability of alloys can also be determined using EELS. A major advantage of EELS over other methods of determining the bonding in solids is that the incident beam can be focused to a small probe, less then 1 nm across, and spectra can be obtained from such small regions. Hence spectra from individual defects can be obtained.

REFERENCES

[1] Howie A and Whelan M J 1961 *Proc. R. Soc.* A **263** 217; 1962 *Proc. R. Soc.* A **267** 206

[2] Egerton R F and Whelan M J 1974 *Phil. Mag.* **30** 739; 1974 *J. Electron Spectrosc.* **3** 232

[3] Smart D J and Humphreys C J 1980 *Inst. Phys. Conf. Ser.* vol 52 (Bristol: Institute of Physics) p 211

[4] Nagata F and Fukuhara A 1967 *Japan. J. Appl. Phys.* **6** 1233

[5] Lally J S, Humphreys C J, Metherell A J F and Fisher R M 1972 *Phil. Mag.* **25** 321

[6] Gjønnes J and Høier R 1971 *Acta Crystallogr.* A **27** 313

[7] Moodie A F, Etheridge J and Humphreys C J 1996 *Acta Crystallogr.* A **52** 596

[8] Hirsch P, Howie A, Nicholson R, Pashley D W and Whelan M J 1977 *Electron Microscopy of Thin Crystals* (Huntington, NY: Krieger)

[9] Etheridge J, Moodie A F and Humphreys C J 1998 *Proc. 14th Int. Conf. on Electron Microscopy (Cancun, 1998)* vol III, ed H A Calderón Benavides and M J Yacamán (Bristol: Institute of Physics) p 737

[10] Aldred P J E and Hart M 1973 *Proc. R. Soc.* A **332** 223, 239

[11] Hewat E A and Humphreys C J 1974 *High Voltage Electron Microscopy* ed P R Swann, C J Humphreys and M J Goringe (London: Academic) p 52

[12] Smart D J and Humphreys C J 1978 *Inst. Phys. Conf. Ser.* vol 41 (Bristol: Institute of Physics) p 145

[13] Spence J C H and Zuo J M 1992 *Electron Microdiffraction* (New York: Plenum)

[14] Saunders M, Bird D M, Zaluzec N J, Burgess W G, Preston A R and Humphreys C J 1995 *Ultramicroscopy* **60** 311

[15] Botton G A, Guo G Y, Temmerman W M and Humphreys C J 1996 *Phys. Rev.* B **54** 1682

[16] Botton G A and Humphreys C J 1997 *Micron* **28** 313

[17] Botton G A, Guo G Y and Humphreys C J 1995 *Inst. Phys. Conf. Ser.* vol 147 (Bristol: Institute of Physics) p 535
Botton G A, Guo G Y, Temmerman W M, Szotek Z, Humphreys C J, Yang Wang, Stocks G M, Nicholson D M C and Shelton W A 1996 *Materials Theory, Simulations and Parallel Algorithms* ed E Kaxiras, J Joannopoulos, R V Vashishta and R K Kalia (Pittsburgh, PA: Materials Research Society) p 567

[18] Botton G A, Guo G Y, Temmerman W M, Szotek Z, Humphreys C J, Yang Wang, Stocks G M, Nicholson D M C and Shelton W A 1998 in preparation

[19] Lipson H and Taylor A 1939 *Proc. R. Soc.* A **173** 232

[20] Cottrell A H 1995 *Intermetallics* **3** 341

[21] Freeman A J, Hong T, Lin W and Xu J-U 1991 *High-Temperature Ordered Intermetallic Alloys IV* ed L A Johnson, D P Pope and J O Stiegler (Pittsburgh, PA: Materials Research Society) p 3

6

INTERPRETATION OF SPATIALLY RESOLVED VALENCE LOSS SPECTRA

A Howie

6.1 ELASTIC AND INELASTIC SCATTERING EFFECTS IN ELECTRON IMAGING

The development and fruitful application of the transmission electron microscope (TEM) in the 1950s generated an urgent need for reliable techniques of image acquisition and interpretation based on electron scattering theory. Two theories were available. The older, more elaborate and quantitative approach was founded on the theory of electron energy loss and stopping power developed for homogeneous assemblies of atoms by Bohr, Bethe, Bloch and Fermi. Here the fast electron is treated as a classical point charge, traversing a medium to whose excitations it transfers energy via its Coulomb interactions and their powerful polarizing and ionizing effects. This theory leads naturally to the idea of mass–thickness contrast in electron images. One of the principal experimental discoveries however was that the electron microscope images of crystals and their defects are much more strongly governed by a diffraction contrast mechanism than by mass thickness effects. Stacking faults, consisting of a pure shear displacement with no local density change whatsoever, afforded one of the earliest and most striking examples of this phenomenon [1]. The appropriate imaging theory [2] then turned out to depend on the elastic scattering of electron waves as described by the Schrödinger equation and first applied to the case of crystals by Bethe and Bloch, but still at that time, in comparison with stopping power theory, a fairly qualitative approach.

Even with the limited data available to electron microscopists in about 1960, this dramatic triumph of elastic, wave scattering theory over stopping power theory was qualified in two respects. Firstly, the ability of the electron beam to induce dislocation movement [3] or indeed, in many cases, visible

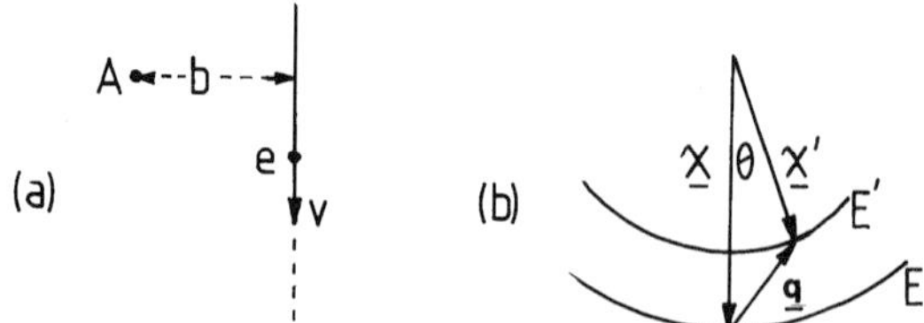

Figure 6.1 Energy and momentum transfer from a fast electron in the real space impact parameter picture (*a*) and in *q* space (*b*).

damage to the specimen signalled the presence of significant inelastic events. Secondly, quantitative agreement between the diffraction contrast theory and the observations was achieved only if so-called anomalous absorption effects were included [4], representing the role of thermal diffuse scattering [5] and other inelastic processes [6] in scattering electrons outside the collection aperture.

Attempts to estimate the appropriate anomalous absorption parameters led to the realization that inelastic scattering mean free paths were frequently considerably less than the thickness of the samples used. A further problem was thus revealed—the need to explain how the various interference effects so characteristic of diffraction contrast imaging could be preserved after inelastic scattering. A mechanism for contrast preservation in crystals, based on the inelastic scattering of Bloch waves, was discovered [7] and found to depend essentially on a comparatively delocalized interaction with small momentum transfer in the inelastic scattering event. Developments in instrumentation, outlined below, soon provided detailed support for these ideas. Images taken with selected energy loss in the conventional transmission microscope [8] showed amazingly close similarity to zero-loss, elastic images. Later on, high resolution scanning transmission electron microscope (STEM) images, using the strongest component of the loss spectrum corresponding to low loss valence excitations, showed the same behaviour [9]. It was thus implied that these processes were rather delocalized on the interatomic distance scale with low momentum transfers and corresponding scattering angles that are small relative to the Bragg angle. The plausibility of this hypothesis was later argued [10] using a simple time-of-flight argument for a fast electron of velocity v passing at distance b from an atom or other excitation centre (see figure 6.1(*a*)). The atom experiences an electromagnetic pulse of characteristic duration $\Delta T = b/v$ and therefore maximum angular frequency $\omega = 2\pi/\Delta T = 2\pi v/b$. The maximum energy loss is thus $\Delta E = \hbar\omega = 2\pi\hbar v/b$. The maximum impact parameter for energy loss ΔE on this picture is therefore given by the expression

$$b = \frac{2\pi\hbar v}{\Delta E}.$$

(6.1)

It is also interesting to note, as shown in figure 6.1(*b*) for a fast electron collision with energy loss ΔE, that $2\pi\hbar b^{-1} = \Delta E/v = \hbar q_{\min}$ is the minimum possible momentum transfer occurring at zero scattering angle. The uncertainty relation implicit here suggests that the larger momentum transfers corresponding to higher angle scattering events will be associated with tighter spatial localization. Equation (6.1) in fact defines the Bohr impact parameter dating back to 1913 which is perhaps a further signal to electron microscopists that it might after all have been useful to read the old stopping power theory literature.

6.2   THE DIELECTRIC MODEL OF VALENCE EXCITATION

For valence excitations ($\Delta E < 50$ eV) generated by 100 keV electrons, equation (6.1) indicates maximum impact parameters in the 5 nm range and thus provides a ready explanation for the complex spectral features generally observed in the low loss region. The fast electron can interact with many valence electrons simultaneously so that collective excitations or plasmons can be generated in addition to the more familiar single electron excitations. A powerful model of valence excitation, capable of including all of these effects, is due to Fermi [11] and makes use of the complex, frequency-dependent dielectric response function $\varepsilon(\omega) = \varepsilon_1(\omega) + i\varepsilon_2(\omega)$. For simplicity here we ignore, initially at least, the dependence of ε on the momentum transfer $\hbar q$. Fermi's theory depends on a simple physical picture in which a charged particle passing through a medium subjects it to an electromagnetic pulse which can be decomposed into components of different frequency ω. Each of these frequency components then induces a polarization of the medium related to $\varepsilon(\omega) - 1$ and this in turn gives rise to an electric field which acts back on the charged particle, slowing it down. In the non-relativistic theory [11–13], the probability of energy loss $\Delta E = \hbar\omega$ is related to the so-called bulk loss function

$$P_b(\omega) = \mathrm{Im}\left\{\frac{-1}{\varepsilon(\omega)}\right\} = \frac{\varepsilon_2(\omega)}{\left(\varepsilon_1(\omega)^2 + \varepsilon_2(\omega)^2\right)}. \tag{6.2}$$

In simple cases, peaks in this loss function can occur because of a maximum in the numerator or a near zero in the denominator. The former can be recognized as single electron losses arising when $\varepsilon_2(\omega)$ (which describes the optical absorption at frequency ω) has a maximum but is still less than $|\varepsilon_1(\omega)|$. On the other hand, when $\varepsilon_1(\omega)$ is zero and $\varepsilon_2(\omega)$ is small, the maximum in the loss function is due to a collective mode or plasmon. The dielectric response function turns out to be a very convenient way of relating energy loss spectra to optical absorption or reflection data. Because ε and $1/\varepsilon$ are both linear response functions, subject to the usual causality constraints, their real and imaginary parts are not independent but connected in each case by Kramers–Kronig relations. The two functions $\varepsilon_1(\omega)$ and $\varepsilon_2(\omega)$ can thus be constructed from the information about $\mathrm{Im}\{-1/\varepsilon(\omega)\}$ obtainable from energy loss spectroscopy. Derivation of such information however depends on obtaining properly normalized energy loss spectra from samples of known thickness with all multiple scattering effects removed by deconvolution techniques. The effects of surface excitations (see below) must also be subtracted either by analysing results as a function of thickness or by computation via an iterative procedure.

For a free electron gas of valence electron density n, the dielectric function is given by the expression

$$\varepsilon = 1 - \frac{\omega_p^2}{\omega(\omega + i\gamma)} \tag{6.3}$$

where $\omega_p = \{ne^2/\varepsilon_0 m\}^{1/2}$ is the free electron plasmon frequency and γ is a damping constant, usually small. A crude but sometimes useful approximation to the dielectric function for an insulator or semiconductor with a bandgap $\hbar\omega_g$ is

$$\varepsilon = 1 - \frac{\omega_p^2}{\{\omega(\omega + i\gamma) - \omega_g^2\}}. \tag{6.4}$$

For small values of damping and $\omega_g \ll \omega_p$, it is easy to show that the loss spectrum will exhibit a collective loss at a resonance frequency of $\omega_r = \{(ne^2/\varepsilon_0 m) + \omega_g^2\}^{1/2}$.

Although Fermi's theory was initially developed to compute the stopping power of matter for fast particles, it attracted considerable interest from condensed matter physicists stimulated by the 1956 Bohm–Pines theory of plasmons. By the time electron microscopists became involved, many aspects of the theory had already been quantitatively tested and developed in the context of broad beam transmission electron spectroscopy measurements on thin films [14]. Bulk plasmon energies and their dispersion relations were measured in a large number of materials. The surface plasmon, a new collective mode predicted by Ritchie [15], was found to be particularly prominent in the loss spectra from very thin films and associated with the excitation function

$$P(\omega) = \mathrm{Im}\left\{ \frac{-1}{(1 + \varepsilon(\omega))} \right\}. \tag{6.5}$$

The relativistic extension of Fermi's theory [12], incorporating radiative losses associated with the Cherenkov mechanism as well as transition radiation, was also explored in these thin film experiments and is discussed further below.

6.3   INSTRUMENTATION FOR SPATIALLY LOCALIZED ENERGY LOSS SPECTROSCOPY

Energy loss spectroscopy combined with high spatial resolution in the TEM became possible in the early 1960s and has been reviewed by Metherell [16].

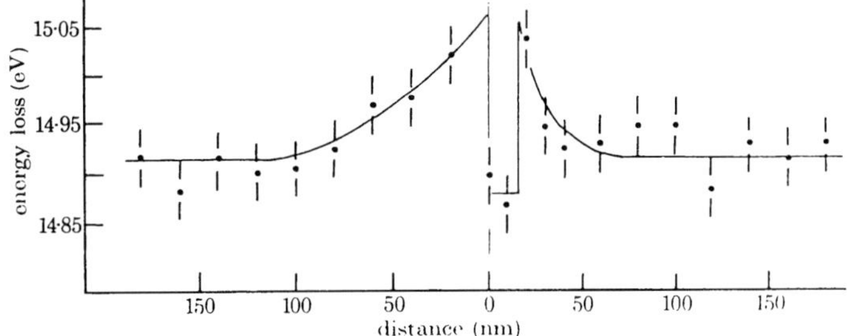

Figure 6.2 Plasmon energy plot near a high angle grain boundary in Al + Mg alloy showing segregation of Mg to the boundary and a denuded zone on either side. (Taken from Cundy *et al* [19], courtesy Royal Society.)

The final magnified image could for instance be superimposed on the entrance slit of a Möllenstedt analyser [17], which, acting as a cylindrical lens, cast an image of the slit on a second image plane with a lateral displacement sensitively dependent on the energy of the electrons involved. One-dimensional specimen features were thus displayed on one axis along the slit with a non-linear display of energy loss at right angles on the other. In the Castaing–Henry system [18], based on a pair of opposing magnetic prisms placed at an intermediate point in the microscope column, a two-dimensional image in any predetermined energy loss could be obtained. The possibility afforded by this new instrumentation for spatially localized energy loss spectroscopy as well as energy-selected imaging was exploited in some early and very impressive studies of segregation in alloys on the nanometre scale [19]. In free electron materials, the bulk plasmon is usually the most prominent feature of the valence excitation region. Precise measurements of the position of this peak in the loss spectrum can thus reveal local changes in valence electron density n arising for example from variations in alloy composition. Figure 6.2 shows an example of segregation of Mg to a grain boundary in an Al + Mg alloy. Right at the grain boundary, the decrease of about 0.05 eV in the plasmon peak position at 15 eV would correspond to an increase in the local Mg concentration there of about 1%. Also clear in the figure is the layer near the boundary which has become slightly denuded of Mg and shows a higher plasmon loss energy. The correlation between plasmon peak position and alloy composition was established empirically with homogeneous calibration samples since it is complicated by the lattice parameter dependence on composition as well as by band structure effects.

Several further applications were made of these plasmon loss fingerprint techniques but the number of suitably simple free-electron-like systems was limited. In transition metal (TM) systems the valence loss spectra are usually much more complicated. The development of the STEM with a sector magnet

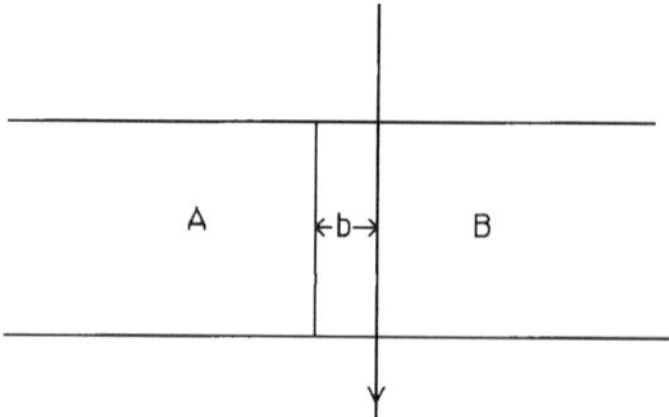

Figure 6.3   Schematic diagram of the case frequently studied in thin film transmission of a planar interface between two dielectric media.

spectrometer [20] and, much later, parallel electron energy loss systems (PEELS) [21] opened the way forward. Practical, spatially localized electron spectroscopy was then extended to the higher loss, core excitation region, where, though relatively weaker, the spectral features are more characteristic of the atoms concerned and thus in general much easier to interpret. Studies of compositional changes on the nanometre scale have thus become more or less routine and detailed studies of the chemical shifts and shape changes above excitation edges are beginning to reveal local changes in electronic structure and bonding. For a wider review of these revolutionary developments in nanoanalysis, which are not described further here, reference should be made to the textbook by Egerton [13]. The acquisition of high quality, spatially localized spectra from valence regions was also greatly facilitated by the STEM instrumentation but it became increasingly clear that their successful interpretation involved the challenge of extending the Fermi theory of dielectric excitation to deal with inhomogeneous media with the complex structures of interest to electron microscopists.

6.4   EXCITATIONS AT PLANAR INTERFACES

A model example for both theory and experiment is the case of a planar interface between two different dielectric media as shown in figure 6.3. The electron beam has been aligned parallel to and at distance b from the interface which itself lies normal to the surfaces of a thin film. For the time being however we ignore the top and bottom surfaces of the film and simply use the expression, quoted by Howie [22] but originally derived by Ritchie, for the probability P_I of energy loss per unit distance along an infinitely extended interface in terms of the momentum transfer $\hbar q_y$ in the plane of the interface and normal to the beam direction.

$$\frac{\mathrm{d}^2 P_I(b, w, q_y)}{\mathrm{d}\hbar\omega\,\mathrm{d}q_y} = \frac{e^2}{4\pi^2 v^2 \varepsilon_0 \hbar^2 v(\omega, q_y)} \left[\mathrm{Im}\left\{ -\frac{1}{\varepsilon_B} \right\} \{1 - \exp\{-2vb\}\} \right.$$

$$\left. + \mathrm{Im}\left\{ \frac{-2}{(\varepsilon_A + \varepsilon_B)} \right\} \exp\{-2vb\} \right] \tag{6.6a}$$

where $v^2 = q_y^2 + (\omega/v)^2$. If the q dependence of the dielectric functions is ignored, the integration over q_y can be carried out up to some (high) cut-off value q_c to obtain the result

$$\frac{dP_I(b, \omega)}{d\hbar\omega} = \left(\frac{e^2}{2\pi^2\varepsilon_0\hbar^2 v^2}\right)\left[\text{Im}\left\{\frac{-1}{\varepsilon_B(\omega)}\right\}\ln\left(\frac{q_c v}{\omega}\right)\right.$$

$$\left. + K_0\left(\frac{2\omega b}{v}\right)\left(\text{Im}\left\{\frac{-2}{(\varepsilon_A(\omega) + \varepsilon_B(\omega))}\right\} - \text{Im}\left(\frac{-1}{\varepsilon_B(\omega)}\right)\right)\right]. \quad (6.6b)$$

Here it is noteworthy that the dependence on distance b from the interface is expressed by the monotonically decreasing Bessel function K_0 whose argument is scaled by the Bohr impact parameter introduced above. At small values of b, K_0 diverges logarithmically (which is related to the need for a cut-off in momentum transfer) but there is a long tail extending beyond $\omega b/v = 1$. This functional dependence had earlier been encountered in the theory of energy loss of slow electrons near surfaces [23]. At large distances from the interface, the only significant contribution originates from the first term containing the bulk loss function for medium B. At smaller values of b however, a new loss function $\text{Im}\{-2/(\varepsilon_A(\omega) + \varepsilon_B(\omega))\}$ appears which is symmetrical in both media and can be recognized as describing an interface loss which is a generalization of the surface plasmon between a medium and the vacuum. The appearance of the interface loss is compensated by an increasing subtraction of the bulk loss known as the *begrenzungs* effect. The *begrenzungs* effect is also observed when an electron travels across an interface. At distances of less than about v/ω from the interface, the appropriate bulk loss function is progressively replaced by the interface function. The Bohr impact parameter thus gives a measure of the film thickness above which bulk losses begin to dominate over surface losses as well as of the spatial localization in spectroscopy when the beam is parallel to an interface. The measure is a very rough one, however, in a complex situation [24], and is strongly affected by the momentum transfers involved in any particular experiment.

The planar interface loss expression has now been tested in a number of cases, including not only internal interfaces in thin films but also the external surfaces of well aligned faceted particles such as MgO smoke cubes [25, 26] whose thickness is accurately known. Figure 6.4 shows for example the bulk and surface loss functions derived by Walls and Howie [27] for an MgO cube from loss spectra obtained for a trajectory penetrating the cube and for a trajectory just outside the cube. Since the complete dielectric function can be effectively recovered from either set of data, their mutual consistency can be checked. Figure 6.4 shows that there is indeed a compromise fit to both sets of energy loss data which is appreciably better than can be achieved with the dielectric function derived from optical measurements.

Equation (6.6) can easily be extended to deal with the grazing incidence conditions employed in reflection high energy diffraction (RHEED) or reflection

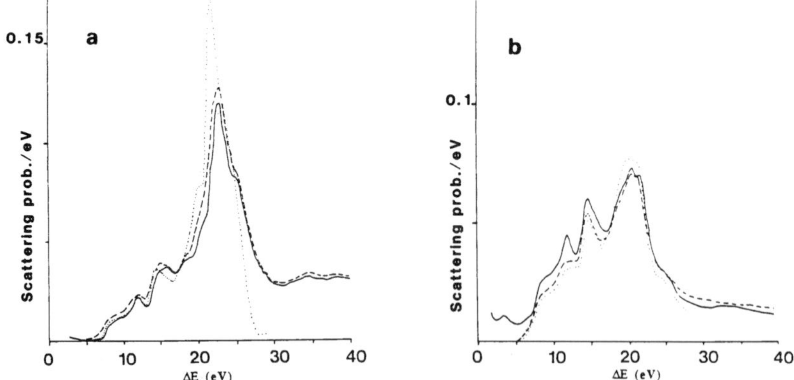

Figure 6.4 Normalized energy loss probabilities (*a*) inside and (*b*) just outside an MgO cube. Measured data (full lines) are compared with optical data (dotted lines) and with compromise dielectric data (broken lines) designed to fit both curves simultaneously. (Taken from Walls and Howie [27], courtesy Elsevier Science.)

electron microscopy (REM) where the electron beam is Bragg reflected by planes parallel to the crystal surface and emerges again after penetrating only a small distance. To a good approximation [28], the trajectory can be broken down into a sequence of short segments each of which is parallel to the surface. For a beam making a small grazing angle θ with the surface and reflecting at the surface we then find that the differential loss probability is given by

$$\frac{\mathrm{d}P(\theta,\omega)}{\mathrm{d}\hbar\omega} = \left(\frac{e^2}{2\pi\varepsilon_0\hbar^2\omega v\theta}\right)\mathrm{Im}\left\{\frac{-1}{(\varepsilon(\omega)+1)}\right\}. \tag{6.7}$$

At steeper angles of incidence θ, the beam begins to penetrate significantly into the crystal so that signs of the bulk loss peak appear. The experimental observations can be quantitatively fitted in considerable detail with the effective penetration depth (which typically turns out to be under 2 nm) as the only adjustable parameter [28].

For a well defined interface plasmon to be observed we require $\varepsilon_A + \varepsilon_B$ to be close to zero. The real parts of ε_A and ε_B should therefore be equal and opposite in sign and the imaginary parts should be small. For free electron materials with plasmon frequencies ω_A and ω_B, the interface plasmon frequency ω_I is given by $\omega_I^2 = (\omega_A^2 + \omega_B^2)/2$ and lies between the two. In the case of semiconductors or insulators, however, we can sometimes obtain an interface plasmon energy close to the band edge. Thus in the Si–SiO$_2$ interface where the two bulk plasmon energies lie at 17 and 26 eV, respectively, the interface plasmon peak is observed [27] at about 8 eV. In principle equation (6.6) can be tested quite thoroughly in such cases since the dielectric functions ε_A and ε_B can be determined from the loss spectra far on either side of the interface before computing the predicted result close to the interface. This approach does indeed give an interface loss at the expected position but considerably more intense than the one observed. The agreement with experiment is significantly improved [27] if the theory is generalized to deal with a sandwich interface produced by inserting at the interface a thin layer of a third material C (SiO in this case). The ultimate generalization of this approach is to periodic planar interfaces such as occur in MBE multilayers. Expressions have been derived [29] for parallel, normal and even oblique incidence on such multilayers.

6.5   MORE COMPLEX INTERFACES

6.5.1   Spherical interfaces

The problem of calculating the valence excitations on a sphere has obvious relevance to the study of small catalyst particles or precipitates in alloys. For a sphere of radius a, an analytical expression can be given [30] for the loss probability for a particle travelling on an aloof trajectory at impact parameter $b > a$:

$$\frac{\mathrm{d}P(b,\omega)}{\mathrm{d}\hbar\omega} = \left(\frac{e^2 a}{\pi^2 \varepsilon_0 \hbar^2 v^2}\right) \sum_{lm} (2 - \delta_{0m}) \frac{P_{lm}(b,\omega)}{(l-m)!(l+m)!} \tag{6.8a}$$

$$P_{lm}(b,\omega) = \mathrm{Im}\{-\gamma_{lm}(\omega)\} \left(\frac{\omega a}{v}\right)^{2l} K_m^2\left(\frac{\omega b}{v}\right). \tag{6.8b}$$

For a sphere of material A embedded in material B,

$$\gamma_{lm} = \frac{(2l+1)}{[l\varepsilon_A(\omega) + (l+1)\varepsilon_B(\omega)]} - \frac{1}{\varepsilon_B(\omega)}. \tag{6.8c}$$

In these equations, the different terms refer to the l, m eigenmodes ($l > |m| > 0$) on a sphere running from the $l = 1$ dipole modes to the large l modes which become indistinguishable from planar modes. The first term in equation (6.8c) gives the characteristic dielectric excitation function for the l, m mode. The second term is the *begrenzungs* effect once more, subtracting the bulk mode to compensate for the increasing excitation of the spherical mode for trajectories close to the sphere. With still more complex expressions for γ_{lm}, equation (6.8) can describe the excitations of a composite sphere with concentric layers of different materials.

For an electron on a penetrating trajectory $b < a$, the same spherical modes will be excited as well as bulk modes of the material A if the sphere is large enough. The loss probability for this case cannot be written in a closed analytical

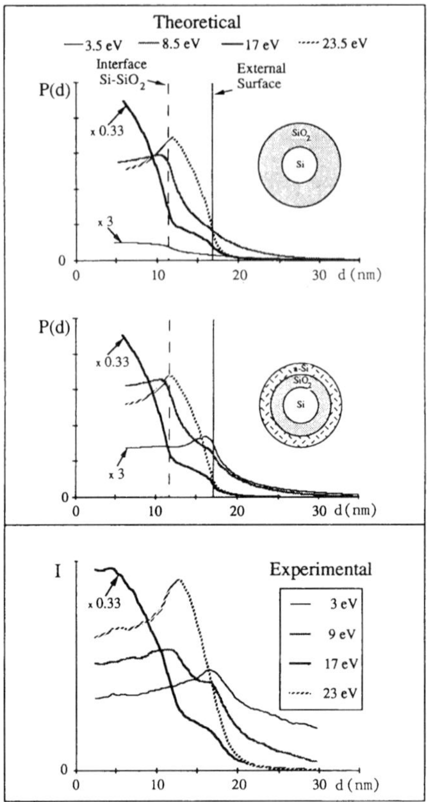

Figure 6.5 Comparison between experimental loss profiles across an oxide-coated Si sphere with theoretical curves for two structural models with and without an outer Si layer. (Taken from Ugarte *et al* [32], courtesy American Physical Society.)

form but some results have been computed [31]. In very small spheres where $\omega a/v \ll 1$, the dipole mode makes the dominant contribution to the loss spectrum from the sphere for both internal and external trajectories but in most practical cases in electron microscopy a considerable number of modes needs to be considered. A particularly detailed experimental study and analysis of the valence losses on small spheres of Si with an SiO coating was made by Ugarte *et al* [32]. They were able to show (see figure 6.5) that a very thin outer layer of Si was also present, perhaps as a result of the reducing action of the electron beam.

In practice of course spheres may be supported on or partially embedded in other materials as well as wholly embedded. The closed analytical expressions for the excitation of a semi-embedded sphere which were derived initially do not appear to be correct but numerical computations have been made for a number of cases of interest [33]. Computations have also been made for hemispheres [34] and for the case of two spheres which are close enough to exhibit coupling effects in their mode excitations [35].

6.5.2 Cylindrical interfaces

Interest in cylindrical interfaces was stimulated by experiments in the STEM where the intensely concentrated electron beam can drill a very narrow hole of diameter as small as 2 nm in various oxide and other ionic materials. Loss spectra can readily be recorded both during the drilling process and afterwards with the beam passing through the completed hole. Analytical expressions for the losses experienced by an electron travelling at various impact parameters parallel to the cylinder axis were successfully derived and compared with the experimental data [36]. The theory has since been generalized to include non-axial trajectories [37]. It can thus also be applied to interpret the very complete spectral data now available from structures such as thin filaments and nanotubes [38].

6.6   MODELLING THE DIELECTRIC RESPONSE FUNCTION

6.6.1 Homogeneous materials

Under the approach just outlined, inhomogeneous samples are usually regarded as composites of homogeneous dielectric regions. Direct measurement in the electron microscope of the necessary input dielectric response functions of the individual component regions may be possible if they are more than a few nanometres in size. Alternatively, if still larger samples are available, high quality optical data extending to the 50 eV loss region and above with the use of synchrotron radiation may be obtainable. For several reasons however, it can be useful to have a more flexible and general model of dielectric response. Such a model can then provide a convenient means of presenting and analysing experimental data as well as a source of trial response functions for the very localized novel phase regions which may occur near interfaces.

One of the simplest approaches to the modelling problem depends on the classical, damped oscillator model which underlies equation (6.4). We can extend this equation to include a number of oscillators with resonance frequencies ω_j, damping constants γ_j and oscillator strengths A_j.

$$\varepsilon(\omega) = 1 - \sum_j \frac{A_j}{\left\{\omega(\omega + i\gamma_j) - \omega_j^2\right\}}. \qquad (6.9)$$

The Kramers–Kronig relations are automatically satisfied by this Lorentz oscillator model and the oscillator resonance frequencies can sometimes be related to energy gaps or transition energies in the electronic band structure. McComb and Howie [39] employed a four oscillator model to fit and analyse their experimental data for the dielectric response of zeolite samples subjected to different dealumination treatment. Although it might seem that a simple measurement of the Al L-shell loss would provide more straightforward monitoring of this process, the Al content is so low that substantial beam damage occurs before an adequate signal can be acquired. By comparison the valence loss spectrum can be collected far more efficiently and appears to vary systematically as dealumination proceeds. The variation in the total valence electron density, obtained from the total oscillator strength $\sum_j A_j$, clearly favours one of the three different models put forward for the structure of the dealuminated site over the others.

The full dielectric function $\varepsilon(\omega, q)$ is strictly needed for the most accurate electron spectroscopy work. Optical measurements provide information only at $q = 0$. At large values of q, the loss function is increasingly dominated by a prominent feature known as the Bethe ridge which peaks at the energy loss $\Delta E = \hbar^2 q^2 / 2m$ corresponding to energy transfer to a free electron initially at rest. The oscillator model provides a useful basis, in conjunction with sum rule constraints and plausible extrapolation schemes, to obtain expressions [40] for the complete loss function $\mathrm{Im}\{-1/\varepsilon(\omega, q)\}$. These extrapolation schemes are clearly not unique however.

Although the oscillator model can provide a rather accurate fit to many features of the experimental data, it is less impressive at dealing with the bandgap and band edge regions of semiconductors and insulators unless a large number of oscillators is employed. In periodic structures, a much more powerful and sophisticated method depends on recognizing that optical excitations are dominated by contributions from a number of critical points in q space where the filled and empty bands run parallel to each other so that the probability of vertical $q = 0$ optical transitions has a singularity. Many, but not all of these points lie on the Brillouin zone boundaries. The characteristic contribution of different points has been identified so that general expressions for the dielectric function can be given [41] and fitting routines are available. Although this is undoubtedly the most powerful approach for the analysis of optical data in periodic structures, it is less clear that it is immediately applicable to electron energy loss data where larger momentum transfer events can be significant and where non-periodic structures are often involved. In recent localized electron spectroscopy studies [42] of the band gap in various semiconductors and insulators, it proved possible to distinguish direct and indirect gap cases depending on whether ε_2 varied as $(\Delta E - E_g)^{1/2}$ or as $(\Delta E - E_g)^{3/2}$ as a function of energy above the bandgap.

The dielectric response function can of course be computed if the electronic band structure is known with sufficient accuracy. In recent years, enormous progress has been made in computing from first principles the optimum atomic structures of various assemblies of atoms including not only simple periodic crystal structures but also atoms in small clusters and atoms at surfaces or interfaces. The atomic structures obtained can then be compared with high resolution EM images. Most of these computations rely on the so-called local density approximation (LDA) in which the complex electron correlation effects are assumed to depend simply on the local electron density [43]. This assumption seems to be much more effective for the computation of the ground state energy and atomic configuration than it is for the excited states which would be required to model the dielectric response. A firm basis exists [44] however for developing other, more ambitious computational models which offer hope of addressing the problem. The computational approach can of course in principle provide complete data on $\varepsilon(\omega, q)$. Periodic continuation methods can also be employed to extend it to derive the dielectric response of inhomogeneous materials.

6.6.2 Effective medium theory

There are a number of instances in which valence loss spectra exhibit features arising from specimen microstructure whose detail is barely or not at all resolvable in the electron microscope image. The dealuminated zeolites mentioned above are one example and another, shown in figure 6.6, is the dispersion of small 'colloidal' Al particles formed in AlF_3 after electron beam irradiation in the STEM. During this irradiation process, as shown in figure 6.7, the main, broad peak at 26 eV characteristic of AlF_3 is reduced and shifted down in energy. At the same time the 10 eV bandgap begins to fill in with the eventual appearance of a broad peak at about 8 eV. A second new and relatively more intense and sharp peak arises at 15 eV. Howie and Walsh [45] explained these effects by averaging the losses which would be experienced by electrons passing along various trajectories through a composite medium consisting of small Al spheres (medium A) embedded at volume fraction f in AlF_3 (medium B). If the response of the composite medium can be represented by an effective dielectric function $\varepsilon_{\text{eff}}(\omega)$, we can then write a simple expression for the effective loss function

$$\text{Im}\left\{\frac{-1}{\varepsilon_{\text{eff}}(\omega)}\right\} = f\left[\text{Im}\left\{\frac{-1}{\varepsilon_A(\omega)}\right\} + g_{\text{int}}\text{Im}\left\{\frac{-3}{(\varepsilon_A(\omega) + 2\varepsilon_B(\omega))}\right\} - \text{Im}\left\{\frac{-1}{\varepsilon_A(w)}\right\}\right]$$

$$+ (1 - f)\left[\text{Im}\left\{\frac{-1}{\varepsilon_B(\omega)}\right\} + g_{\text{ext}}\text{Im}\left\{\frac{-3}{(\varepsilon_A(\omega) + 2\varepsilon_B(\omega))}\right\}\right.$$

$$\left. - \text{Im}\left\{\frac{-1}{\varepsilon_B(\omega)}\right\}\right]. \tag{6.10}$$

In this expression, the main terms with factors f and $(1 - f)$, respectively, describe the contributions from trajectory segments inside and outside the particles. In each case there is firstly an appropriate bulk loss contribution but then, depending how close the trajectory passes to the interface between the sphere and the AlF_3, this contribution is modified by a boundary excitation with a concominant *begrenzungs* effect reduction in the bulk loss. On the assumption

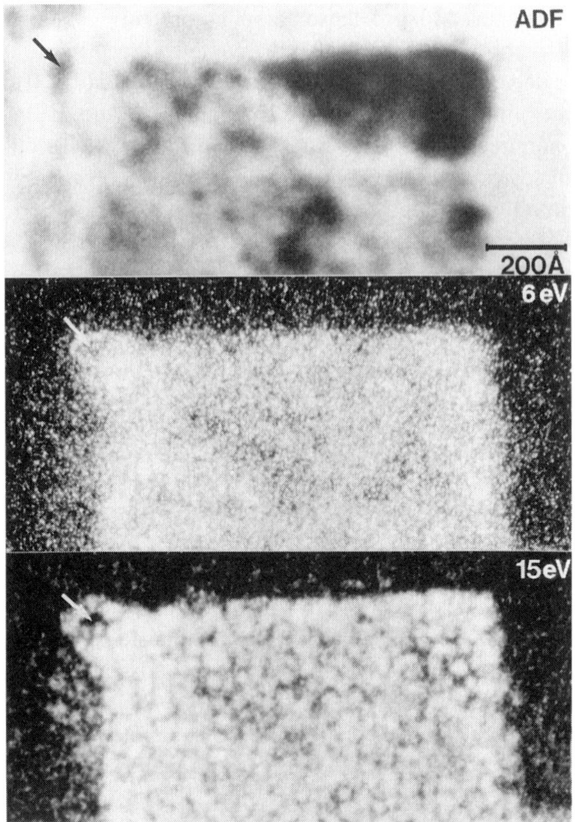

Figure 6.6   STEM annular dark field image (top), 6 eV loss image (middle) and 15 eV loss image (bottom) of the 'colloidal' dispersion of small Al particles formed in AlF_3 as a result of e-beam irradiation. (Taken from Howie and Walsh [45], courtesy Editions de Physique.)

that the spheres are very small, the dipole interface mode has been assumed here, but for larger spheres, as discussed in section 6.5.1 above, an excitation function more similar to the planar function might be appropriate.

The parameter g_{int} in the above equation determines the relative importance of the interface to bulk excitation inside the sphere. The planar excitation theory described by equation (6.6) indicates that interface excitations could be generated by trajectories at a typical distance v/ω from the interface. We therefore expect $g_{int} = 1$ for spheres of radius a much less than v/ω. By referring to the computations of Echenique *et al* [31], Howie and Walsh [45] proposed the empirical relation

$$g_{int} = \frac{1}{(1 + 3a\omega/v)}. \tag{6.11}$$

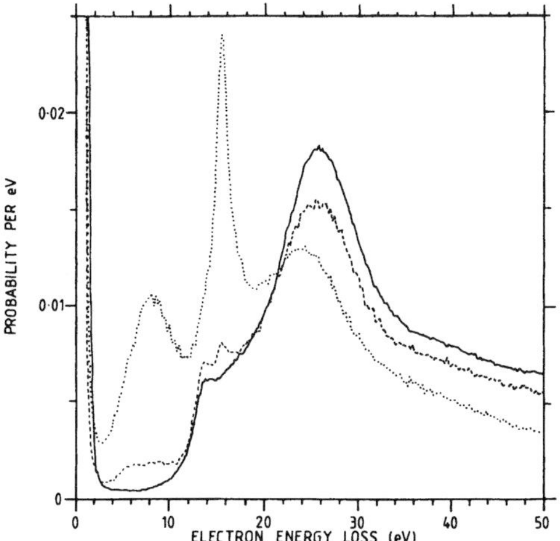

Figure 6.7 Valence loss spectra obtained from AlF$_3$ during the course of e-beam damage (for commentary see text). (Taken from Howie and Walsh [45], courtesy Editions de Physique.)

It has been pointed out [45] that, in the case of very small spheres where $g_{\text{int}} = 1$, equation (6.10) bears a close relation to the well known Maxwell Garnett expression for the response of an effective medium. Indeed, provided we take $g = 2f/(1 + 2f)$, the two expressions are identical. In this limit however, the theory does not give a very satisfactory explanation of the results shown in figure 6.7. The bulk term from AlF$_3$ and the dipole interface term (at about 7 eV) are present but the peak at 15 eV, presumably corresponding to the bulk loss term for Al (material A), is completely missing. In its more general form for larger spheres, equation (6.10) represents a considerable improvement on the Maxwell Garnett theory and, written in terms of the characteristic bulk and interface electron excitations, has a far more transparent appearance.

Equation (6.10) is in fact consistent with a much more general expression which can be deduced from the theories [46] of Bergman and of Milton for the effective response of a composite medium

$$\frac{1}{\varepsilon_{\text{eff}}(\omega)} = \sum_j \frac{A_j}{\{a_j \varepsilon_A(\omega) + (1 - a_j)\varepsilon_B(\omega)\}}. \tag{6.12}$$

Here the excitation modes are determined by the parameters α_j ($0 \leqslant \alpha_j \leqslant 1$) which like the mode strengths A_j depend on the geometry of the composite of which the effective medium is composed. Two sum rules impose a further restriction:

$$\sum_j A_j = 1 \qquad \sum_j A_j \alpha_j = f. \tag{6.13}$$

For optical excitations, the A_j and α_j quantities can all be regarded as constants. With the larger momentum transfers involved in electron excitation, very roughly characterized by ω/v, the mode strengths A_j can be functions of $a\omega/v$ where a is a characteristic dimension of the structure such as the average sphere radius in the example described here. It is interesting to note that the *begrenzungs* effect built into equation (6.10) guarantees the first of these sum rules and the second rule is satisfied provided $g_{\text{ext}} = 2g_{\text{int}} f/(1 - f)$.

A fresh light was thrown on equation (6.10) by Barrera and Fuchs [47] who considered the excitations of an assembly of interacting spheres by a plane wave of momentum $\hbar q$. The interactions were simplified by an averaging procedure allowing for the distribution of the spheres. They were able to provide for this specific case a rigorous derivation of equation (6.10) together with explicit forms for the quantities g_{int} and g_{ext} as functions of q. After integrating over a range of q appropriate to the collection aperture in the typical electron beam experiment, it was found [48] that equation (6.11) provides a reasonably good fit to the dependence on ω and a. This work makes more manifest the fact that equation (6.10) is really an approximate extension of the Maxwell Garnett theory to the case of finite momentum transfer.

It remains an interesting challenge to find a method of readily computing the parameters A_j and α_j for specific geometries of the components in the composite. Zeolite structures constitute a particularly intriguing example of this problem [49]. Despite the very large difference in mean density, the main plasmon peak in the loss spectrum is only slightly shifted down relative to those of silica or quartz, which can perhaps be taken to be typical of the material in the walls of the zeolite channels and cages. Attempts to model the dielectric function of a zeolite by applying Clausius–Mosotti theory (essentially Maxwell Garnett theory) to a fine-scale mixture of silica and vacuum [49] were not very successful. At this comparatively crude level it appears that there is little consensus on the correct approach to problems of this kind even after discussion for more than a century. Perhaps equation (6.12) could provide the basis for a new perspective.

6.7   RELATIVISTIC EFFECTS

Since electron velocities in electron microscopy can easily exceed half the velocity of light, the non-relativistic theory of dielectric excitation can often be inadequate. The use of Poisson's equation in a quasi-static approximation to calculate at each frequency ω the fields induced by a moving charge has to be replaced with Maxwell's equations. The non-instantaneous propagation of the electromagnetic pulse from this moving source to a point in the polarizable medium and then back to the source to slow it down gives rise to retardation effects. New radiative energy loss mechanisms also become possible, including Cherenkov radiation when the fast electron velocity exceeds the velocity of light in the dielectric medium and transition radiation when the fast electron crosses from one dielectric medium to another. Expressed as a function of momentum transfer $\hbar q$ as well as energy transfer $\hbar\omega$, the relativistic bulk loss function is [12]

$$P_b(\omega, q) = \mathrm{Im}\left(-\frac{\left(1 - \varepsilon(\omega, q)\beta^2\right)}{v(\omega, q)\,\varepsilon(\omega, q)}\right) \tag{6.14}$$

where $\beta = v/c$ and $v^2 = q^2 + (\omega/v)^2 - \left(\varepsilon(\omega, q)\omega^2/c^2\right)$. The vanishing of this latter quantity is evidently linked to the Cherenkov condition for direct energy loss via the excitation of a transverse EM wave in the medium propagating at an angle $\theta = \arctan(qv/\omega)$ to the electron beam direction.

With the development of thin film, broad beam transmission electron spectroscopy, the relativistic theory was extended by Kliewer and Fuchs and by Kroger [50] to cover the case of a fast electron traversing a thin dielectric slab. Kroger's formula for the energy loss includes not only the bulk electronic and Cherenkov loss contributions but also transition radiation losses as well as the contributions from surface modes, which can be radiative or non-radiative outside the slab depending on whether their wave vector q parallel to the surface is less than or greater than ω/c. The formula for Cherenkov losses was successfully tested many years ago [51] with data from semiconductor thin films where $\varepsilon > 4$ in the region below the bandgap. Much more recently, with interest reviving in the topic, some direct measurements of both forward and backward Cherenkov emission from thin films have been made in the electron microscope [52]. Although the radiation would not be so easily detectable, Cherenkov radiation trapped by internal reflection inside a thin film can also be excited and contribute to the energy loss [53].

A relativistic generalization of equation $(6.6a)$ for incidence parallel to a dielectric interface has been obtained [54] and can be written in a form showing the dependence of the excitations on impact parameter b as well as the operation of the *begrenzungs* effect at each value of q.

$$\frac{\mathrm{d}^2 P_I(b, \omega, q_y)}{\mathrm{d}\hbar\omega\,\mathrm{d}q_y} = \frac{e^2}{4\pi^2\varepsilon_0\hbar^2 v^2}\left[\mathrm{Im}\left\{-\frac{\left(1 - \varepsilon_B\beta^2\right)}{v_B\varepsilon_B}\right\}(1 - \exp(-2v_B b))\right.$$

$$\left. +\mathrm{Im}\left\{\frac{2}{(v_B\varepsilon_A + v_A\varepsilon_B)} - \frac{2\beta^2}{(v_A + v_B)}\right\}\exp(-2v_B b)\right] \tag{6.15}$$

where $v_{A,B}^2 = q^2 + (\omega/v)^2 - \left(\varepsilon_{A,B}(\omega, q)\omega^2/c^2\right)$.

The application of this expression to experimental data for the Si–SiO$_2$ interface has been studied in some detail [55]. In particular it was found essential to include this q-dependent, relativistic form for the interface excitation function in order to account for the observed shift of the interface loss peak from near 8 eV at small impact parameters down to about 7 eV at impact parameters of 5 nm. The analysis in addition confirmed the suggestion [27] that the interface appears to contain a 1 nm layer of other material with a response function like SiO. Detailed study of equation (6.15) also reveals that, with the beam travelling in medium B, Cherenkov radiation losses can occur when its velocity exceeds the velocity of light in *either* medium A *or* in medium B. So far however, a direct demonstration of such losses with the beam outside the medium concerned has not been obtained. In the infinitely extended Si/SiO_2 interface the Cherenkov condition is reached (for Si) at electron energies of 100 keV for most energy losses below about 4.5 eV. The condition will however be more stringent for an interface in a thin film.

In cylinders, the retarded mode structures (radiative and non-radiative) have been worked out as well as the energy loss probability for electrons travelling with different impact parameters parallel to the axis [56]. A fully analytical expression for the relativistic losses due to a dielectric sphere has yet to be found (see, however [69]) but a good deal can be learnt from computations of specific cases. In this case *all* of the modes exhibit both electronic and radiative damping [57]. However, the radiative damping is strongest for the low l modes which are usually excited only on very small spheres where $\omega a / v \ll 1$. On larger spheres, such modes are probably more strongly excited when the velocity of light in the sphere is less than that of the fast electron (even if the latter does not pass through the sphere).

6.8 NUMERICAL SIMULATION OF LOSS SPECTRA BY BOUNDARY CHARGE METHODS

6.8.1 Eigenmodes for bodies of arbitrary shape

The samples studied in the electron microscope often have a structure too complex to be adequately approximated by one of the simple shapes for which analytical solutions of the valence excitations are available. Fortunately however an electrostatic method, originally introduced by Maxwell [58] to estimate capacitances, can be adapted for the purpose. The method depends on finding a self-consistent distribution of boundary charges on all the dielectric interfaces and was used by Fuchs [59] to find the eigenmodes of a small cube and later reformulated by Ouyang and Isaacson [60] for a body of arbitrary shape. In the presence of an external potential $\phi^{ext}(s, \omega)$ with frequency ω generated at an interface point s by a passing fast electron, the self-consistent condition for the interface charge density $\sigma(s, \omega)$ takes the form [61]

$$\Lambda(\omega)\sigma(s, \omega) = 4\pi\varepsilon_0 n_s \cdot \nabla\phi^{ext}(s, \omega) + \int d^2 s'\, \sigma(s', \omega) n_s \cdot \nabla \frac{1}{|s - s'|}$$

$$(6.16a)$$

where

$$\Lambda(\omega) = 2\pi \frac{\varepsilon_B(\omega) + \varepsilon_A(\omega)}{\varepsilon_B(\omega) - \varepsilon_A(\omega)}.$$

$$(6.16b)$$

Here n_s is a unit vector at the point s on the interface directed normal to the interface from medium A to medium B. A standard route to solving equations of this type is firstly to find the eigenmodes $\sigma^j(s)$ and eigenvalues $2\pi\lambda_j$ of the equation in the absence of the external potential

$$2\pi\lambda_j\sigma^j(s) = \int d^2s'\,\sigma^j(s')\,n_s \cdot \nabla\frac{1}{|s-s'|}. \tag{6.17}$$

With standard computer packages and an adequate number of sampling points (typically about 100) on the interface, the eigenmodes can readily be computed and, in the case of simple shapes like a sphere, checked against the known results. For any combination of the two dielectric materials, equation (6.16b) can then be employed, setting Λ equal to each $2\pi\lambda_j$ value in turn to determine the mode frequencies which will in general be complex because of damping. It may be observed that interface charge interactions in a plane do not produce any electric field component normal to the plane so that all the eigenvalues $2\pi\lambda_j$ are zero in that case. Although the integrand in equation (6.16a) is not symmetrical under interchange of s and s', the eigenvalues $2\pi\lambda_j$ can in general be shown to be real [60] and furthermore the eigenfunctions $\sigma^j(s)$ form a complete basis set, satisfying an orthogonality property

$$\int d^2s \int d^2s'\frac{\sigma^i(s)\,\sigma^j(s')^*}{|s-s'|} = \delta_{ij}. \tag{6.18}$$

With the aid of this equation, the complete solution of equation (6.16a) can be obtained as a sum over eigenmodes with excitation amplitudes C_j

$$\sigma(s,\omega) = \sum_j C_j(\omega)\,\sigma^j(s) \tag{6.19a}$$

where

$$C_j = \frac{4\pi\varepsilon_0}{\Lambda - 2\pi\lambda_j}\int d^2s \int d^2s'\frac{\sigma^j(s')^*n_s \cdot \nabla\phi^{\text{ext}}(s,\omega)}{|s-s'|}. \tag{6.19b}$$

When the fast electron trajectory lies entirely in medium B, this equation becomes

$$C_j(x,y,\omega) = G_j(\omega)\int d^3r \int d^2s \int d^2s'\,\nabla\left\{\frac{\rho^{\text{ext}}(r,\omega)}{|s-r|}\right\} \cdot \frac{n_s\sigma^j(s')^*}{|s-s'|} \tag{6.20a}$$

where

$$G_j(\omega) = -\left\{\frac{1}{2\pi(1+\lambda_j)}\right\}\left\{\frac{2}{(1-\lambda_j)\varepsilon_B + (1+\lambda_j)\varepsilon_A} - \frac{1}{\varepsilon_B}\right\}. \tag{6.20b}$$

We may recognize here yet again the typical form of an interface mode with the corresponding subtraction of the bulk mode via the *begrenzungs* effect. The factor G_j in the mode excitation, depending on the interface geometry and the dielectric functions of the media concerned, determines the form of the mode frequency response. In addition to the values of λ_j characteristic of a plane ($\lambda_j = 0$) or the lth mode on a sphere ($\lambda_j = -1/(2l+1)$), it appears likely that in some cases it should be possible to associate particular values of λ_j with local features such as edges or corners, although the precise sharpness of the wedge or corner is important [61, 62]. The remaining factor in equation (6.20*a*) for the mode excitation contains the dependence on the incident electron position $r = r_0(x, y) + vt$, i.e. impact parameter components x, y of the incident electron. In line with previous analytical results, we would expect these to enter in scaled form $x\omega/v$ and $y\omega/v$ as the arguments of some function. Clearly the eigenmodes with largest values of $\left|\sigma^j(s)\right|$ near to the fast electron trajectory will be those which are most strongly excited.

For any particular interface charge distribution $\sigma(s, \omega)$, the stopping power and energy loss probability function $P(x, y, \omega)$ can be directly expressed by evaluating the electric field component E_z which it generates at the position of the fast electron. Taking $s = (s_\rho, s_z)$ where s_ρ is the projection of s in the (x, y) plane normal to the fast electron trajectory, we obtain

$$\frac{dP(x, y, \omega)}{d\hbar\omega} = \frac{e}{\pi\varepsilon_0\hbar^2 v} \int d^2s \, K_0\left(\frac{\omega|r_0 - s_\rho|}{v}\right) \text{Im}\left\{\sigma(s, \omega) \exp\left(\frac{-i\omega s_z}{v}\right)\right\}.$$

$$(6.21)$$

The appearance here of an apparent phase shift depending on the coordinate s_z is in fact compensated by an opposite phase factor in the generation of $\sigma(s, \omega)$ by the fast electron. As was previously found for the particular case of a sphere in equations (6.8), the probability of energy loss can in general be written as a sum of contributions from different eigenmodes if we use equations (6.19) and (6.20) to express $\sigma(s, \omega)$ in terms of its mode components $\sigma^j(s)$. Clearly those modes with largest values of $\left|\sigma^j(s)\right|$ close to the fast electron trajectory, which we have already seen will be the most strongly excited, will also make the largest contribution to the energy loss. The impact parameter thus enters twice in determining the significance of any particular mode for energy loss.

6.8.2 Energy loss spectra in complex geometries

With the aid of standard computer packages, the theory just described can be employed to compute non-relativistic energy loss spectra near dielectric interfaces with a complex geometry. It is particularly efficient for systems with some symmetry along the beam direction, e.g. cylindrical symmetry, such as a dielectric wedge aligned with the beam, or axial symmetry, such as coupled spheres aligned with the beam [61]. In such cases the interface charge density

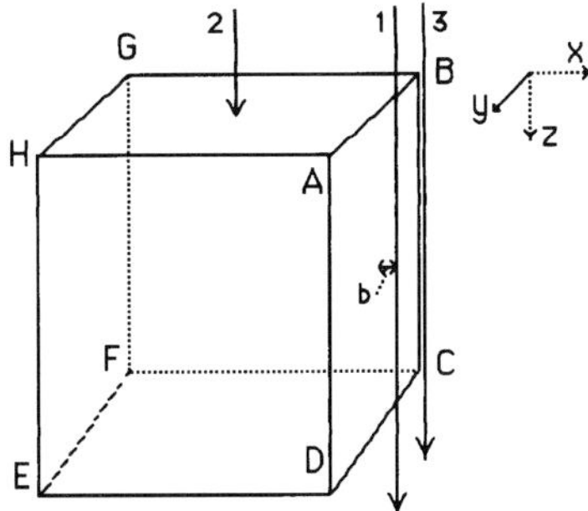

Figure 6.8   Schematic diagram of various possible trajectories in or near an MgO smoke cube.

can often be adequately sampled with as few as 20 points s. In more general situations, a few hundred points could be required. The method has been successfully employed [63] to assess the importance of edge effects in the loss spectra from MgO cubes with various possible trajectories shown in figure 6.8. Figure 6.9 shows experimental results and simulations for parallel trajectories passing near a face centre or near an edge of a long MgO square cylinder. The relatively greater weight of the region below 15 eV due to the excitation of edge modes is apparent. These computations thus effectively include, respectively, the nearest face or the nearest two faces and edge of the cube but ignore the effects of the top and bottom surfaces and edges or corners. By taking a perpendicular trajectory near a square cylinder, the top and bottom contributions can be included at the expense of ignoring the side contributions which should be a good approximation for face-centred trajectories near all but the smallest cubes. The results for this case are shown in figure 6.10 and indicate that the correction to the simple infinite plane model of equation (6.4) due to top and bottom surface or edge effects is mostly apparent below 15 eV and begins to be significant for cube sizes of less than 50 nm.

It is clear that, with more computing effort, realistic simulations could be carried out for more complex shapes such as a complete cube. The theory can also be applied to situations where more than two dielectric media are present, e.g. a sphere of material A supported in vacuum on a thin film of material B [64]. As indicated by equation (6.16b) however, Λ must then be regarded as a function of s when used in equation (6.16a) so that a single computation of the eigenmodes and eigenvalues is no longer sufficient but a self-consistent solution has to be found for the particular dielectric combination involved.

6.8.3   Approximation methods

The numerical simulations so far carried out indicate that significant corrections to the results of the simplest analytical models may often be necessary but that it

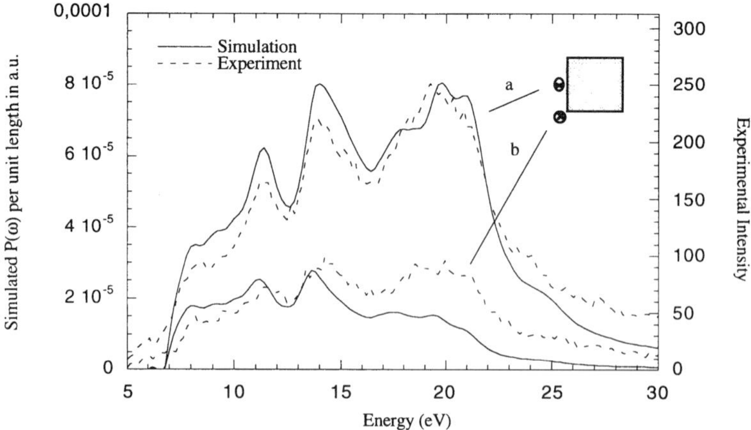

Figure 6.9 Comparison between normalized loss probabilities (broken lines) for trajectories 1 and 3 (shown in figure 6.8) and simulated data (full lines). (Taken from Aizpurua *et al* [63].)

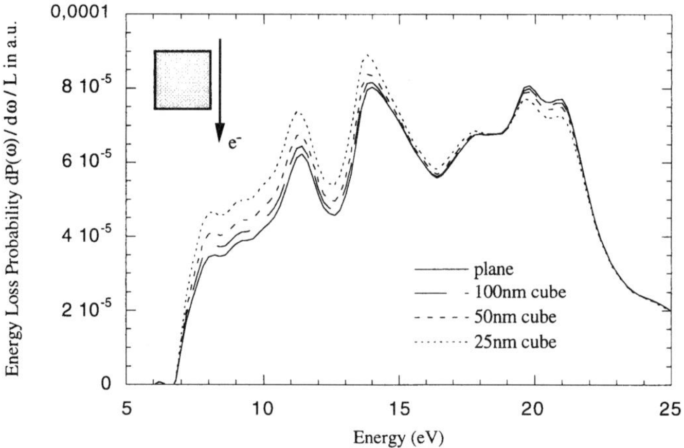

Figure 6.10 Computed loss probabilities per unit length of cube size L for trajectories of type 1 shown in figure 6.8. The failure of the curves for different cube sizes L to superimpose is due to the effects of the top and bottom faces and edges. (Taken from Aizpurua *et al* [63].)

may be quite demanding to solve every problem by first computing all the normal modes. An obvious alternative approach is to try to solve equations (6.16*a*) and (6.21) by some iterative procedure. We can then take account of the fact that the simplest analytical approximation may be a reasonable starting point and that further corrections to the stopping power arising from some iterative approach of the interface charge density towards self-consistency in equation (6.16*a*) will be most significant for interface points s closest to the fast electron trajectory.

The rather simple scheme described here does not always converge but does illustrate some useful points. If we denote the nth approximation to the interface charge distribution by $\sigma^{(n)}(s, \omega)$, the next approximation is given by

$$\Lambda(w)\sigma^{(n+1)}(s, \omega) = 4\pi\varepsilon_0 n_s \cdot \nabla\phi^{\text{ext}}(s, \omega) + \int d^2 s'\, \sigma^{(n)}(s', \omega)n_s \cdot \nabla\frac{1}{|s - s'|}$$

$$(6.22a)$$

and

$$\Lambda(\omega)\sigma^{(1)}(s, \omega) = 4\pi\varepsilon_0 n_s \cdot \nabla\phi^{\text{ext}}(s, \omega)$$

$$= \frac{e\omega}{\pi\varepsilon_B v^2\eta} \exp\left\{\frac{i\omega s_z}{v}\right\}\{n_s \cdot \eta K_1(\eta) + i\eta n_s \cdot n_z K_0(\eta)\} \quad (6.22b)$$

where $\eta = \omega(s_\rho - r_0)/v$. In general it appears that the second term in this expression does not give any contribution to the stopping power, at least in the first approximation, when integrated over a closed interface and can therefore be ignored. For a planar interface, as already stated, the integral in equation (6.22*a*) is zero and equation (6.22*b*) gives a result which is exact. Substitution of equation (6.22*b*) into equation (6.21) indeed yields the usual expression given in equation (6.6*b*) for the loss probability function near a plane if we make use of a result proved by Dr Garcia de Abajo (private communication)

$$\int_0^\infty \frac{dx}{\left(x^2 + b^2\right)^{1/2}} K_0\left\{\frac{\omega\left(x^2 + b^2\right)^{1/2}}{v}\right\} K_1\left\{\frac{\omega\left(x^2 + b^2\right)^{1/2}}{v}\right\} = \frac{\pi v}{\omega a} K_0\left\{\frac{2b\omega}{v}\right\}.$$

$$(6.23)$$

The MgO cube shown in figure 6.8 would probably be an excellent example to test out the usefulness of this approximation scheme. A preliminary study for the external trajectory 1 suggests that, in the first approximation, the contributions to the loss probability from the top and bottom surfaces cancel, leaving only the parallel face ABCD closest to the fast electron. The difference between this face and the infinite plane considered in equation (6.6*b*) arises in this approximation because of its finite lateral extent in the y direction normal to the beam and could be readily estimated by changing the upper limit of integration in equation (6.23). The finite extent of the face ABCD in the beam direction is correctly included if the differential stopping power given by equation (6.21) is multiplied by the length of the face. The three other side faces will, in the first approximation, all give contributions opposite in sign to the face closest to the beam but obviously considerably smaller in magnitude because of their greater distance. In the next approximation, the additional charges induced on the main face by the first order charges on the top and bottom surfaces will still be opposite in sign and cancel one another. However, the first order charges on the face ABCD will induce second order charges of the same sign on the top and bottom faces and this is probably the main second order effect.

The important practical case previously shown in figure 6.3 of an interface normal to the surfaces of a thin film could be a particularly fruitful object for the approximation scheme. As remarked above, the exact eigenvalue approach is more complicated to apply when more than two media (including the vacuum region) are present. Because the beam traverses different dielectric regions, some of the cancellation effects noted for the top and bottom surfaces with the wholly external trajectory will no longer apply.

The convergence of the approximation scheme can be conveniently investigated if the charge distribution at various steps is expanded in terms of the eigenmode distributions

$$\sigma^{(n)}(s, \omega) = \sum_j A_j^{(n)} \sigma^j(s). \tag{6.24}$$

Using equations (6.22a), (6.17) and (6.18) we then find

$$\Lambda(\omega) A_j^{(n+1)} = \Lambda(\omega) A_j^{(1)} + 2\pi \lambda_j A_j^{(n)} \tag{6.25}$$

which leads to the result

$$A_j^{(n+1)} = \frac{A_j^{(1)}}{(1-\xi)}\left(1 - \xi^n\right) \qquad \text{where} \quad \xi = \frac{2\pi \lambda_j}{\Lambda}. \tag{6.26}$$

The condition $|2\pi \lambda_j / \Lambda| < 1$, which is required for this process to converge satisfactorily to the correct result, cannot unfortunately be guaranteed if $|\Lambda/2\pi| < 1$. Professor R Fuchs (private communication) has however devised a much more powerful approximation scheme directly for the stopping power itself and based on the recursion method.

6.8.4   Relativistic simulations

The considerably greater complexity of analytical solutions in the relativistic regime, and the correspondingly small number of those available, makes it highly attractive to explore the possibility of extending the numerical approach just described to include retardation effects. One might hope that it would be possible to model the response of a dielectric composite with the aid of a suitable distribution of interface charges and currents, interacting self-consistently with one another as well as with the fast electron. This possibility is currently being actively explored in joint work between the author and Dr J Garcia de Abajo.

Maxwell's equations in an inhomogeneous dielectric can indeed be written in a form where the terms in $\nabla\varepsilon$ appearing at interfaces emerge as current or charge sources. However, the Coulomb potential which was previously employed to govern the interactions between charges at points s and s' in equation (6.16a), now has to be written in retarded form to take account of the velocity of electromagnetic waves in the different intervening media

$$\phi_{\mathrm{ret}}(s - s') = \frac{1}{|s - s'|} \exp\left\{ \frac{-\mathrm{i}\omega|s - s'|\sqrt{\varepsilon}}{c} \right\}. \tag{6.27}$$

This retarded Green function is discontinuous at the dielectric interfaces and additional interface charges and currents are required in compensation. Preliminary results [70] are very encouraging. It can be shown directly that the method correctly reproduces the eigenmodes (including radiative damping effects) for planes, cylinders and spheres as well as energy loss spectra which agree with those already obtained by other means. More difficult problems such as coupled spheres are now being tackled. It will then be necessary to compare the method in speed and accuracy with some of the other methods which have already been developed for solving problems involving the generation or scattering of radiation in inhomogeneous media. These include the Purcell–Pennypacker method in which the optical properties of a small particle are simulated by a volume distribution of internal dipoles with appropriate frequency response [65] and the method introduced by Pendry [66] for numerically solving Maxwell's equations in a periodic medium.

6.9   CONCLUSIONS AND FUTURE PROSPECTS

The non-relativistic theory of valence losses is now in a reasonably satisfactory state. Indeed the valence loss spectrum for a fast electron traversing a complicated but well defined structure can now be calculated with more quantitative accuracy than is yet possible for the elastic scattering, at least in the context of HREM images (though possibly not convergent beam electron diffraction (CBED) patterns). The use of the classical, point charge model, which may well underpin this achievement, was shown [67] to be valid in most situations where virtually all of the inelastic scattering is collected. In a more complete picture, the elastic and inelastic scattering should clearly be handled together quantum mechanically. Such an approach might be necessary for instance if electron spectroscopy became more significant in weak beam imaging or in annular dark-field imaging. Kohl and Rose [68] made a model computation of the excitation of a point dipole scattering from a STEM probe wave function into plane waves while Rivacoba *et al* [33] have given a general, self-energy formalism valid for any target geometry and dielectric response. Since most elastic electron scattering computations in complex structures are handled by periodic continuation and the slice method, inelastic scattering between coupled plane waves or perhaps Bloch waves would seem to be the most natural format. The concepts of intraband and interband inelastic Bloch wave transitions, developed many years ago in simple crystal structures, may therefore have to be revisited in the more demanding situation of artificially large unit cells. Some of the density matrix methods recently adapted to study the gradual loss of coherence between Bloch waves as a result of inelastic scattering [71] could then prove useful.

The long history of stopping power theory compared with the wave theory of high energy electron scattering is an index of its wider significance in physics and related branches of science. Although the wave theory has obvious connections to electron band structure theory as well as to the dynamical theory of x-ray diffraction, it is essentially a 'niche' theory, highly adapted to the needs of electron microscopists and little understood by workers in other fields. On the other hand, stopping power theory retains its interest in nuclear physics applications and in radiation damage studies. Via the dielectric response function, it has direct links to optical properties, photonic band structures and van der Waals interactions as well as chemical and electronic behaviour. The development of stopping power theory in a form suitable for the study of excitations in local regions of complex structures is thus of potential importance in a wide range of fields outside electron microscopy, particularly if relativistic effects can be included.

ACKNOWLEDGMENTS

I am grateful to P M Echenique, R Fuchs and J Garcia de Abajo for many valuable discussions and to the colleagues mentioned in the text for permission to use illustrations.

REFERENCES

[1]   Whelan M J and Hirsch P B 1957 *Phil. Mag.* **2** 1121, 1303
[2]   Hirsch P B, Howie A, Nicholson R B, Pashley D W and Whelan M J
         1965 *Electron Microscopy of Thin Crystals* (London: Butterworth)
[3]   Whelan M J, Hirsch P B, Horne M J and Bollmann W 1957 *Proc. R. Soc.*
         A **240** 524
[4]   Hashimoto H, Howie A and Whelan M J 1960 *Phil. Mag.* **5** 967
[5]   Whelan M J 1965 *J. Appl. Phys.* **36** 2103
         Hall C R and Hirsch P B 1965 *Proc. R. Soc.* A **286** 158
[6]   Whelan M J 1965 *J. Appl. Phys.* **36** 2099
[7]   Howie A 1963 *Proc. R. Soc.* A **271** 268
[8]   Castaing R 1969 *Z. Angew. Phys.* **27** 171
[9]   Craven A J and Colliex C J 1977 *J. Microsc. Spectrosc. Electron.* **1** 511

[10] Howie A 1979 *J. Microsc.* **117** 11

[11] Fermi E 1940 *Phys. Rev.* **57** 485

[12] Landau L D and Lifshitz E M 1960 *Electrodynamics of Continuous Media* (New York: Pergamon)

[13] Egerton R F 1986/1996 *EELS in the Electron Microscope* (New York: Plenum)

[14] Raether H 1965 *Springer Tracts in Modern Physics* vol 38 (Berlin: Springer) pp 84–157

[15] Ritchie R H 1957 *Phys. Rev.* **106** 874

[16] Metherell A J F 1971 *Advances in Optical and Electron Microscopy* vol 4, ed V E Cosslett and R Barer (London: Academic) pp 263–360

[17] Möllenstedt G 1949 *Optik* **5** 499

[18] Castaing R and Henry L 1962 *C. R. Acad. Sci., Paris* **255** 76

[19] Cundy S L, Metherell A J F, Whelan M J, Unwin P N T and Nicholson R B 1968 *Proc. R. Soc.* A **307** 267

Cundy S L, Metherell A J F and Whelan M J 1968 *Phil. Mag.* **17** 141

[20] Crewe A V, Isaacson M A and Johnson D 1971 *Rev. Sci. Instrum.* **42** 411

[21] Egerton R F 1981 *Proc. 39th EMSA Meeting* (Baton Rouge, FL: Claitors) p 368

Shuman H and Kruit P 1985 *Rev. Sci. Instrum.* **56** 231

Krivanek O L, Ahn C C and Keeney R B 1987 *Ultramicroscopy* **22** 103

McMullan D, Rodenburg J M, Murooka Y and McGibbon A J 1990 *Inst. Phys. Conf. Ser.* vol 98 (Bristol: Institute of Physics) p 55

[22] Howie A 1983 *Ultramicroscopy* **11** 141

[23] Echenique P M and Pendry J B 1975 *J. Phys. C: Solid State Phys.* **8** 2936

[24] Batson P E 1992 *Ultramicroscopy* **47** 133

Muller D A and Silcox J 1995 *Ultramicroscopy* **59** 195

[25] Marks L D 1982 *Solid State Commun.* **43** 727

[26] Howie A and Milne R H 1985 *Ultramicroscopy* **18** 427

[27] Walls M G and Howie A 1989 *Ultramicroscopy* **28** 40

[28] Bleloch A L, Howie A, Milne R H and Walls M G 1989 *Ultramicroscopy* **29** 175

[29] Bolton J P R and Chen M 1995 *J. Phys.: Condens. Matter* **7** 3373, 3389

[30] Ferrell T L and Echenique P M 1985 *Phys. Rev. Lett.* **55** 1526

Echenique P M, Howie A and Wheatley D J 1987 *Phil. Mag.* B **56** 335

[31] Echenique P M, Bausells J and Rivacoba A 1987 *Phys. Rev.* B **35** 1521

[32] Ugarte D, Colliex C and Trebbia P 1992 *Phys. Rev.* B **45** 4332

[33] Rivacoba A, Zabala N and Echenique P M 1992 *Phys. Rev. Lett.* **69** 3362

[34] Aizpurua J, Rivacoba A and Apell S P 1996 *Phys. Rev.* B **54** 2901

[35] Batson P E 1982 *Phys. Rev. Lett.* **49** 936

Schmeits M and Dambly L 1991 *Phys. Rev.* B **44** 12706

Zabala N, Rivacoba A and Echenique P M 1997 *Phys. Rev.* B **56** 7623

[36] Walsh C A 1989 *Phil. Mag.* A **59** 227

Zabala N, Rivacoba A and Echenique P M 1989 *Surf. Sci.* **209** 465

[37] Rivacoba A, Apell P and Echenique P M 1995 *Nucl. Instrum. Methods* B **96** 465

[38] Cullen S L, Botton G, Kirkland A I, Brown P D and Humphreys C J 1993 *Inst. Phys. Conf. Ser.* vol 138 (Bristol: Institute of Physics) p 79

[39] McComb D W and Howie A 1990 *Ultramicroscopy* **34** 84

[40] Ashley J C 1988 *J. Electron Spectrosc. Relat. Phenom.* **46** 199
Ritchie R H and Howie A 1977 *Phil. Mag.* **36** 463

[41] Loughin S, French R, De Noyer L K, Ching W-Y and Xu Y-N 1996 *J. Phys. D: Appl. Phys.* **29** 1740

[42] Rafferty B K and Brown L M 1998 *Phys. Rev.* B **58** 10 326

[43] Payne M C P, Teter M P, Allan D C, Arias T A and Joannopoulos J D 1992 *Rev. Mod. Phys.* **64** 1045

[44] Hedin L 1965 *Phys. Rev.* **139** A796

[45] Howie A and Walsh C A 1991 *Microsc. Microanal. Microstruct.* **2** 171

[46] Bergman D J 1978 *Phys. Rep.* **43** 377
Milton G W 1981 *J. Appl. Phys.* **52** 5286

[47] Barrera R G and Fuchs R 1995 *Phys. Rev.* B **52** 3256

[48] Fuchs R, Barrera R G and Carrillo J L 1996 *Phys. Rev.* B **54** 12 824

[49] McComb D W and Howie A 1995 *Nucl. Instrum. Methods* B **96** 569

[50] Kliewer K L and Fuchs R 1966 *Phys. Rev.* **144** 495; *Phys. Rev.* **150** 573
Kroger E 1968 *Z. Phys.* **216** 115

[51] Daniels J, von Festenberg C, Raether H and Zeppenfeld K 1970 *Springer Tracts in Modern Physics* vol 54 (Berlin: Springer) p 77

[52] Yamamoto N, Sugiyama H and Toda A 1996 *Proc. R. Soc.* A **452** 2279

[53] Chen C H and Silcox J 1975 *Phys. Rev. Lett.* **35** 390

[54] Garcia Molina R, Gras Marti A, Howie A and Ritchie R H 1985 *J. Phys. C: Solid State Phys.* **18** 5335

[55] Moreau P, Brun N, Walsh C A, Colliex C and Howie A 1997 *Phys. Rev.* B **56** 6774

[56] Walsh C A 1991 *Phil. Mag.* **63** 1063

[57] Fuchs R and Kliewer K L 1968 *J. Opt. Soc. Am.* **58** 319

[58] Maxwell J C (ed) 1921 *The Scientific Papers of the Hon. Henry Cavendish FRS* (Cambridge: Cambridge University Press)

[59] Fuchs R 1975 *Phys. Rev.* B **11** 1732

[60] Ouyang F and Isaacson M 1989 *Phil. Mag.* B **60** 481

[61] Garcia de Abajo F J and Aizpurua J 1997 *Phys. Rev.* B **56** 15 873

[62] Dobrzynski L and Maradudin A A 1972 *Phys. Rev.* B **6** 3810
Davis L C 1976 *Phys. Rev.* B **14** 5523

[63] Aizpurua J, Rafferty B, Garcia de Abajo F J and Howie A 1997 *Inst. Phys. Conf. Ser.* 153 (Bristol: Institute of Physics) pp 277–80

[64] Ouyang F and Isaacson M 1989 *Ultramicroscopy* **31** 345

[65] Purcell D M and Pennypacker C R 1973 *Astrophys. J.* **186** 705
Draine B T and Flatau P J 1994 *J. Opt. Soc. Am.* **11** 1491
Rupping R 1996 *Z. Phys.* D **36** 69

[66]   Pendry J B and MacKinnon 1992 *Phys. Rev. Lett.* **69** 2772
       Pendry J B and Martin-Moreno L 1994 *Phys. Rev.* B **50** 5062
       Pendry J B, Holden A J, Stewart W J and Youngs I 1996 *Phys. Rev. Lett.* **76** 4773
       Garcia Vidal F C and Pendry J B 1996 *Phys. Rev. Lett.* **77** 1163
[67]   Ritchie R H and Howie A 1988 *Phil. Mag.* A **58** 753
[68]   Kohl H and Rose H 1985 *Adv. Electron. Electron Phys.* **65** 173
[69]   Garcia de Abajo F J 1999 *Phys. Rev.* B **59** 3095
[70]   Garcia de Abajo F J and Howie A 1998 *Phys. Rev. Lett.* **80** 5180
[71]   Dudarev S L, Peng L-M and Whelan M J 1993 *Phys. Rev.* B **48** 13 408

7

IS MOLECULAR IMAGING POSSIBLE?

*J C H Spence, X Huang, U Weierstall,
D Taylor and K Taylor*

7.1   INTRODUCTION

In an important paper written at the time of the first observation of atoms by electron microscopy [1], Breedlove and Trammel [2] showed that, because of radiation damage, the direct imaging of individual organic molecules by electron or x-ray microscopy was not possible. Most of our knowledge of structural biology has thus come from crystallographic techniques, which provide periodically averaged information. Since an understanding of molecular shape is the key to understanding molecular interactions and function, and since many important biomolecules cannot be crystallized (or may change shape when doing so), this is an important conclusion.

Recent work in scanning tunnelling and atomic force microscopy has shown dramatic results for molecular imaging; however image interpretation and reproducibility remain severe problems. In view of this work, perhaps a better title for this paper would be 'Is molecular imaging possible in transmission?'. Professor M J Whelan's life-long interest in, and contributions to, the theory of inelastic electron scattering are crucial to answering this question, as we shall see.

The analysis of Breedlove and Trammel was based on the ratio $R = \sigma_i/\sigma_e$ of inelastic to elastic cross sections. Elastic scattering contributes to image formation, while inelastic scattering may cause damage. For light atoms, R falls to about unity as the electron beam kinetic energy E_k falls to 100 eV ($\lambda = 0.12$ nm), below which the diffraction limit may limit the spatial resolution of images. (Although ionization cross sections increase with decreasing beam energy, elastic cross sections increase more rapidly.) σ_i is about 1 Å^2 at 100 eV. Assuming that a fraction $0.5 < \varepsilon < 1$ of ionizations lead to dissociation, and

108

that at least one elastic scattering event is needed per atom to form an image, they therefore conclude that imaging of undamaged molecules is impossible. At 1 kV, for example, they estimate three inelastic interactions from hydrocarbons for every four elastic ones. Values of R are much greater for x-ray illumination at all energies, due to the photoelectric effect.

This analysis has been criticized on many detailed grounds (e.g. that cross sections are not relevant to coherent image formation, the neglect of substrate interactions etc), but generally it has stood the test of time. More recent analyses, supporting this conclusion, have been based on the Rose criteria for detecting signals amongst noise [3], and a different figure of merit $Q = (\sigma_i/\sigma_e)\Delta E$, the amount of energy deposited by inelastic events per elastic event, has been proposed and applied to a comparison of electron beams, x-rays and neutrons for imaging [4]. Perhaps closest to achieving the goal of molecular imaging is the work of Fujiyoshi and Uyeda [5], who imaged a single unsupported strand of DNA drawn across a hole by TEM. The published image, which clearly shows the helical pitch of DNA, has been averaged five times along the axis. Very recent STM and AFM work also appears very promising as problems of reproducibility and image interpretation are slowly being tackled. The essential difficulty, for molecules in solution, is holding them down rigidly during STM imaging (see [6] for a review). The advantages of the point-projection microscope (PPM) to be described here over STM work are the following. (i) Simpler image interpretation (very similar to LEED) due to the higher energy used. (ii) Better decoupling of tip effects. (iii) Transmission PPM images will reveal internal details of large molecules, rather than their surface topography. (iv) The support of molecules on and within solid xenon thin films (as in matrix isolation spectroscopy) or within lipids will prevent their motion. (v) PPM images of molecules also imaged by STM will provide additional information to aid the difficult process of STM image interpretation.

The purpose of this paper is to re-visit the arguments of Breedlove and Trammel in the light of more recent knowledge, and to discuss the latest experimental results from our low voltage point-projection microscope (PPM). Specifically, we ask whether molecular imaging might not be possible at about 7 eV ($\lambda = 0.46$ nm) with subnanometre resolution, by taking advantage of several effects: (i) the continued decline in R to values below 0.01 which has been measured below 20 eV in hydrocarbons [7]; (ii) the various beam-energy damage thresholds, below the lowest of which (at about 8 eV for hydrocarbons) molecular dissociation becomes negligible [8–10] and (iii) the increase in inelastic mean free path λ_i which is known to occur as electron energy decreases below about 20 eV [7], leading to increased penetration of the beam with decreasing voltage. 'Universal curves' $\lambda_i(E_k)$ show a minimum around 30 eV [11]—our aim has been to build a point-projection electron microscope which operates at energies below this minimum and below the threshold for damage. Measurements give $2 < \lambda_i < 10$ nm for $50 < E_k < 5$ eV for light elements, but larger values of λ_i can occur at certain energies, as discussed below.

The plan of this chapter is as follows. Firstly, we give a very brief qualitative review of the main elastic and inelastic scattering processes which occur at low energy, and define the quasi-elastic region. We then summarize what is known from various techniques regarding damage mechanisms in this range. Imaging of particular molecules is then discussed. Finally, we discuss our recent experiments on point-projection imaging of lipids at about 100 V.

The requirements for imaging insulating organic films in transmission at low voltage are: (i) an absence of charging effects; (ii) an inelastic mean free path greater than the sample thickness, or a method of embedding molecules in an electron-transparent film; (iii) sufficient forward elastic scattering for image formation; (iv) weaker inelastic scattering involving processes which do not lead to permanent atomic reorganization and (v) a method for making self-supporting films thinner than λ_i. Previous work on polymer fibres imaged by PPM has been described by Binh [12].

7.2   ELECTRON SCATTERING AT LOW ENERGIES

Several distinct mechanisms affect the elastic and inelastic scattering of electrons in thin insulating films at energies below 50 eV. Most of our understanding in this field of hot electron transport in insulators [13] comes from the techniques of photoemission (UPS), low energy electron transmission (LEET), high resolution electron energy loss spectroscopy (HREELS) and low energy electron diffraction (LEED). The dominant effects are phonon scattering, electron–electron scattering, coherent elastic scattering (including reflection from both surfaces of a thin film and Bragg scattering within it) and defect scattering. Some of these processes may be kinematically forbidden in certain energy ranges, with useful consequences.

1. Below a certain beam kinetic energy E_k^c, inelastic electron–electron scattering cannot occur. (We measure E_k upward from the vacuum level of the sample— this is not exactly the accelerating potential if contact potentials exist between sample and electron gun [14].) Figure 7.1 illustrates the definition of this quasi-elastic scattering range.

   As shown, the beam electron loses the maximum energy $E_{\max} = \phi_A + E_k$ possible if it falls to the bottom of the empty conduction band, but if $E_{\max} < E_g$, where E_g is the band-gap, there is insufficient energy to promote a crystal electron across the gap. Thus if

$$E_k + \phi_A < E_g$$

   or

$$E_k = E_k^c < E_g - \phi_A \tag{7.1}$$

   no electronic energy loss processes are possible. This results in a large inelastic mean free path λ_i. For the inert gas solids, where most experimental

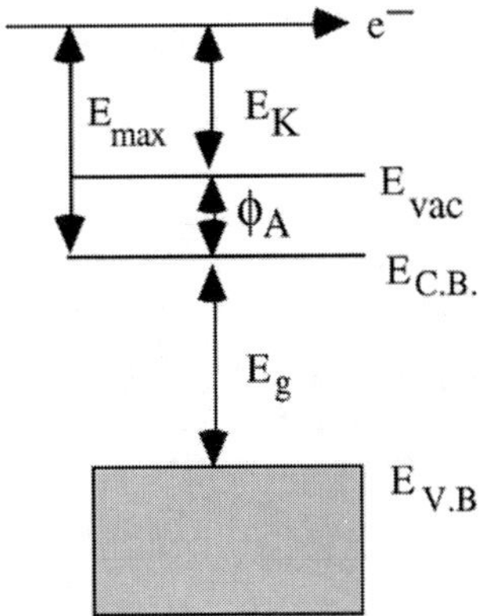

Figure 7.1 The quasi-electric scattering region. No electronic excitations are possible if $E_k + \phi_A < E_g$. The band-gap E_g, the electron affinity ϕ_A and the beam kinetic energy E_k are all indicated.

research has been done, E_g is typically 10 eV and ϕ_A about 1 eV, so that these materials are fairly transparent to electrons at low energy, with penetration limited by diffraction, defects, excitons and acoustic phonon scattering (since no optical phonons are supported). For solid xenon at 55 K, $E_g = 9.3$ eV, $\phi_A = 0.5$ eV and $\lambda_i > 100$ nm for $E_k < 8$ eV [15, 16]. For solid argon, $E_k^c = 11.6$ eV. For LiF, $E_g = 12$ eV and $\phi_A = 1$, while for MgO, $E_g = 8.7$ and $\phi_A = 1$. These and similar materials with large band-gaps and small affinities are sought for use as photocathodes, since their large effective 'escape depth' produces large secondary electron yields [17, 18]. The high intensity of SEM images of insulators compared with metals has also been explained in this way [19].

2. At the smallest thicknesses (a few monolayers), quantum size effects are seen due to interference between waves reflected between the entrance and exit faces of the film [20]. This produces oscillations in the transmission with thickness and energy.

3. At larger thicknesses (more than a few nanometres) the electronic band-structure of the film develops in the direction of the surface normal for empty states above the vacuum level occupied by the beam electron. As is well known in the LEED literature, the band-gaps which result correspond to energies for which reflectivity is high [21] (since there are no states in the film), and satisfy the LEED Bragg condition for back-reflection. Transmission is a minimum. These reflection conditions occurring within the quasi-elastic energy range are essential for image formation in the low energy electron microscope (LEEM). Between the band-gaps transmission is maximized.

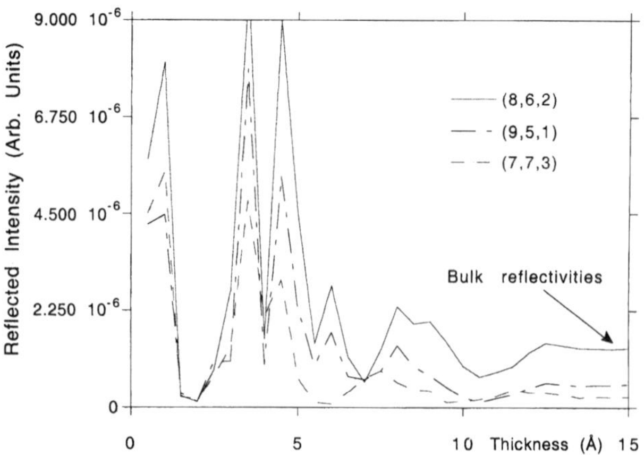

Figure 7.2 Intensity of 250 eV electron Bragg beams reflected from a thin crystal of gold [110] as a function of thickness. The reflectivities increase to about 0.1 at very low voltage, where transmissivity is approximately complementary to reflectivity in the quasi-elastic range.

The oscillations in transmission and reflection due to these coherent elastic multiple scattering processes have again been studied in great detail in thin films of inert gas solids [16, 22]. They are shown in figure 7.2 from our own transmission 'LEED' Bloch-wave calculations for the point-projection microscope [23]. For a thin film, in the quasi-elastic region, the oscillations in transmitted intensity are approximately complementary to the LEED back-reflected beams.

4. The range of inelastically scattered electrons is fixed by the inelastic mean free path λ_i (see, e.g., [24] for a review). The simplest theoretical approach is based on the dielectric constant $\varepsilon(\omega, q)$, since λ_i^{-1} is related to an integral over the imaginary part of $\varepsilon(\omega, q)^{-1}$, which may be obtained from optical measurements [25]. The q dependence of $\varepsilon(\omega, q)$ must be modelled. Experiments measure an attenuation length, which gives only a lower limit for λ_i, since elastic backscattering and finite collection angles both reduce the signal detected. Not all the inelastic scattering events contributing to λ_i cause damage. In the quasi-elastic region $E_k < E_k^c$, where phonon scattering dominates, the differing influences of optical and acoustic phonons may be observed. For example, in hydrocarbons [7], the minimum in $\lambda_i(E_k)$ occurs at around 50 eV, rising at lower energies as electronic transitions switch off and acoustic phonon scattering takes over with a mean free path λ_{ac} proportional to $E_k^{-\alpha}$ [26]. In the inert gas solids, λ_{ac} continues to increase to the lowest energies measurable. Hydrocarbons, however, also support optical (polar) modes [27], with a mean free path λ_{opt} proportional to E_k. These become

important below about 3 eV, so that the curve $\lambda_i(E_k)$ reaches a maximum near this energy, falling away rapidly toward zero kinetic energy.

7.3   ELECTRON BEAM DAMAGE AT LOW ENERGIES

As beam energy decreases, the number of channels available for energy transfer decreases. Inner shell excitations, for example, which are believed to make an important contribution to damage in TEM because of the large amount of energy deposited per interaction [28, 29], cannot occur at energies below 20 eV. It is also known that the unsaturated, pi-bonded aromatic hydrocarbons are more resistant to radiation damage than the aliphatics. For molecules and thin films adsorbed on a substrate, the dominant damage mechanism at low energy is electron-stimulated desorption (ESD) or DIET (desorption of ions by electronic transitions). This depends on the electronic relaxation time being long—comparable to the nuclear relaxation time—so that energy transfer to the nuclear system can occur and atomic displacement result. The important ESD mechanisms include dissociative attachment (DA) and, at higher energies, dipolar dissociation (DD) [30]. The DA process consists of resonant electron capture by a neutral molecule, followed by direct dissociation of the anion into a neutral molecule and a stable anion. The energy of the DA resonance is highly material dependent, but for saturated and unsaturated hydrocarbon films including benzene the resonance occurs at about 10 eV [8]. Below the threshold at 8 eV, ESD is negligible; at 12 eV the cross section σ_d is about 5×10^{-2} Å^2 [9]. (This is measured using a mass spectrometer to monitor directly the production of anions as a function of the energy of an incident monochromatic electron beam.) The same material (benzene) has been the subject of spot-fading experiments in LEED at higher energy (20 eV) [31]— here a cross section of 6 Å^2 was deduced, assuming damage efficiency $\varepsilon = 1$. It is suggested that, apart from the difference in beam energy and assumptions for ε, this difference may arise because LEED spot fading measurements are sensitive to small atomic displacements of surface layer atoms, whereas mass spectroscopy measures ejected anions. No LEED spot-fading measurements appear to have been done on continuous organic films in the quasi-elastic region—these would be of great interest. Stray magnetic fields, charging and the diffraction limit make these difficult experiments.

We may now apply the analysis of Breedlove and Trammel to the transmission imaging of a thin film of benzene at 7 eV. LEET studies show that, as for other hydrocarbons, elastic scattering dominates both vibrational and electronic scattering in the 1–10 eV range [32] (unlike the reverse situation in TEM). However, it is the cross section σ_d for molecular dissociation which is most relevant. Taking $\sigma_d = 10^{-4}$ Å^2 at 7 eV [9] and $\sigma_e > 1$ Å^2 [32] gives $\varepsilon R = \sigma_d/\sigma_e < 10^{-4}$, so that imaging of individual molecules now becomes possible in principle. The cross section for any other damage processes is less than that for ESD. More than 10 000 elastic scattering events then contribute to the image per dissociation. We note also that linear phase contrast imaging is more sensitive than this ratio of cross sections suggests, since they refer to second order (dark-field image) terms [33]. If the damage criterion of Henderson [4] is used (the amount of energy Q deposited by inelastic events per elastic event), taking $\Delta E = 7$ eV, we have $Q = (\sigma_i/\sigma_e)\Delta E = 1.9$ eV. (Here the LEET measurements of [34] are used to estimate approximate cross sections. They find probabilities of 0.79, 0.1 and 0.11 for elastic, phonon and electronic scattering in benzene at 10 eV. Phonon scattering is assumed not to contribute to σ_e.) This very favourable value of Q can be compared with typical values of $Q = 60$ eV for HREM and $Q = 400$ keV for soft x-ray imaging. Since phonon scattering involves unresolvably small energy losses but large scattering angles it can be expected to contribute a diffuse background to PPM images. In many molecular crystals, the dominant lowest energy excitations are local polar modes rather than crystal modes, so that the thermal scattering is isotropic.

Further information on damage at low energies relevant to the PPM can be obtained from the sparse LEED work that has been done on continuous organic films. In one such study [31] at energies above 20 eV, the times to extinction of LEED patterns were found to vary from five seconds (for n-paraffins) to 30 s for naphthalene, to hours for tryptophan and phthalocyanines, at a beam current of 0.5 A m^{-2}. This corresponds to a dose of 0.15 e$^-$ Å^{-2} for the n-paraffins and 0.9 e$^-$ Å^{-2} for naphthalene (or 0.5 e$^-$ Å^{-2} for benzene), which is similar to the doses used to study the purple membrane in TEM. Evidently in these materials, above the ESD threshold, the loss of excitation channels as beam energy is decreased is just compensated for by the increase in inelastic cross section. Hexagonal ice and ammonia are found to be very stable, possibly due to the fact that their decomposition products are volatile, and so do not affect the LEED pattern. For ammonia (naphthalene) these workers find that extinction times are increased from 60 to 600 s (from 10 to 30 s) as the beam energy is decreased from 60 to 40 V (40 to 25 V).

For coronene on MoS$_2$(0001), LEED patterns have been recorded at 80 eV and a dose of 0.5 e$^-$ Å^{-2} [35]. These appear to show much stronger diffraction spots than those obtainable by TEM.

We note that the energy proposed here for imaging hydrocarbons (7 eV) corresponds to the ultra-violet region of the optical spectrum, so that electron damage effects at this energy might be expected to be very roughly comparable to ultra-violet damage.

7.4   EXPERIMENTAL RESULTS

Since self-supporting films of benzene could not easily be made, the purpose of these experiments was to measure the critical dose for the thinnest organic film which readily forms self-supporting films, at the lowest possible voltage. DPPC C16 lipid films (16 carbon atoms and 2 nm thick) may be formed over holes in holey carbon grids. Lipids are important materials in the formation of cell membranes, and are amongst the most radiation sensitive materials. They do not charge up in the PPM. Images were recorded on our PPM instrument at 110 eV. In these experiments, due to the high current density used, the films were destroyed at around 100 V, before it was possible to focus and obtain images at lower voltage.

Dipalmitoyl-L-a-phosphatidyl choline (Sigma) was dissolved in chloroform to a concentration of 0.5 mg lipid ml^{-1}. Monolayers were prepared on Teflon blocks with wells 5 mm in diameter and 1 mm deep milled into them. 1 μg of lipid solution was applied to the surface of distilled water at 4 °C and the chloroform allowed to evaporate in air. Holey grids were prepared by the Fukami–Adachi method [36]. These were placed carbon side down on the lipid monolayers. After lifting the grid and lipid, excess water was carefully removed by touching filter paper to the edges of the grid. The grid was then allowed to air dry.

The microscope has been described in detail elsewhere [37, 38], and has demonstrated sub-nanometre resolution. A tungsten nanotip field-emitter at a potential of about -100 V is placed a distance z_1 (a few thousand ångströms) from a grounded sample, which acts as the anode. A shadow image is formed on a channel plate, distance $z_2 = 14$ cm away, with a magnification $M = z_2/z_1$. Such an image can be shown [39] to be identical to a TEM image (apart from aberrations), which is out of focus by z_1. (We therefore refer to tip approach as 'focusing'.) These images may be considered as in-line electron holograms, and reconstructed by a variety of methods [40], including the Gerchberg–Saxton algorithm [41] and methods based on the transport of intensity equations [42]. The image resolution is approximately equal to the electron source size and to the finest Fresnel fringe. CCD camera exposure times were 0.1 seconds with about 1 nA beam current at 110 V. The tip voltage needed to sustain field emission steadily decreases as z_1 decreases, causing increased magnification and decreasing the focus defect z_1.

Figure 7.3 shows point-projection electron shadow images of lipid films, recorded on our PPM instrument, at 110 eV electron beam energy. Two pictures were taken with a time interval of one second between them. The field emission current was approximately 1 nA. Holes were seen to open up in the sample over a period of a few seconds. The damaged region cannot be carbon film because many separate experiments with holey carbon films under the same conditions clearly show that carbon films are not affected by the beam. Since the illuminated area was estimated to be 1 μm, the electron dose needed to destroy the lipid film at 110 V was approximately 1 nC $\mu m^{-2} = 60$ e^{-} $Å^{-2}$.

In figure 7.3(a), the profile of a holey carbon edge can be seen in the bottom part of the figure. Compared with the second picture, it seems that the region to the left of that edge may have been covered by lipid film at the time when the first picture was taken. Then the lipid covering the hole was removed before the

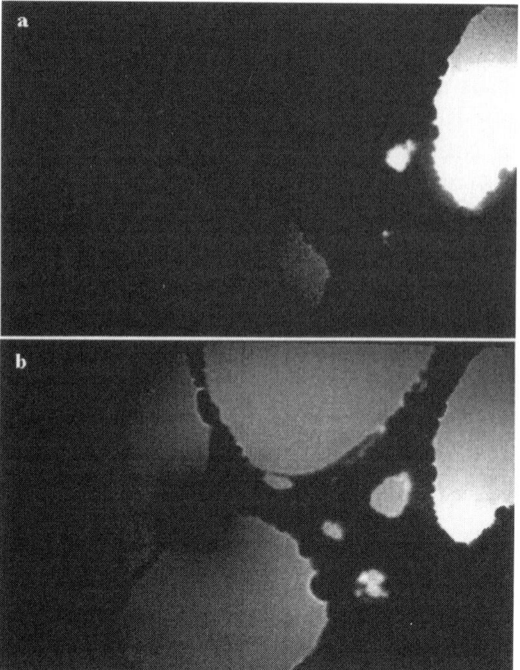

Figure 7.3 Damage to ultra-thin lipid film supported on holey carbon grid. Picture (*a*) is taken about one second before (*b*). Tip bias voltage is -110 V, field emission current ~ 1 nA. The electron dose to cause such damage is estimated to be about 60 e^- $\mathring{A}^{-2}$. In picture (*a*), the observation of a profile of holey carbon edge at the bottom of the picture may indicate transmission through the lipid film, which is removed by electron beam before picture (*b*) is taken.

second image was recorded. This suggests (but does not conclusively establish) some limited transmission of electrons through the film. At this energy the inelastic mean free path is close to its minimum value. Variations in emission intensity from the tip are also evident—the PPM image is superimposed on a field emission image of the emitter.

As the tip approaches the sample (keeping the field-emission current constant by reducing tip voltage) the illuminated sample area, which is proportional to the square of the tip-to-sample distance, decreases. This leads to a rapid increase in current density for the same beam current. Most tips start to emit when far from the sample at about 200 V, with a correspondingly low current density at the sample and stable images. If the voltage is reduced to 100 V (by decreasing z_1 to a magnification of perhaps 30 000) the sample is fairly rapidly destroyed by the high current density. Attempts were also made to use STM servo-control electronics during tip approach, and stable currents were obtainable from conducting samples at low voltages by this method. As discussed below, in future work we plan to use a chevron two-stage channel plate, and so operate at picoamp beam currents rather than nanoamps.

Sample charging problems have not been encountered for lipid samples. This is consistent with LEED results, which find no charging for insulating materials on a conducting substrate at sample thicknesses smaller than 10 nm. Severe charging effects were observed in the PPM from other large band-gap insulators, such as CaF_2, possibly due to the large thickness of sample exposed to the beam. Charging was also observed for polymer balls extending from the edges of the holes of holey carbon film.

In summary, although PPM images of amorphous carbon fibres have been published at energies as low as 7 V [43], we have not been able to image lipids below the ESD threshold. The most important problem is the destruction of the film during focusing (tip approach) due to the rapid increase in current density as voltage decreases. Thus low dose and specialized focusing techniques may be needed for PPM, as in TEM. It appears also that these very thin films may be drawn onto the tip at voltages below about 40 V ($z_1 < 100$ nm), causing tip failure. In addition, magnification becomes excessive at very low voltage (a single Fresnel fringe may fill the 75 mm diameter screen) and the image contains too few pixels to be useful. Solutions to these problems are suggested below.

In a second series of experiments we attempted to obtain images of radiation insensitive material at the lowest voltage possible, and to compare this with a TEM image of the same object. Figure 7.4(a) shows a point projection image of an amorphous fibre from a holey carbon film, illuminated by a 40 eV electron beam ($\lambda = 0.2$ nm). TEM images of the same fibre were also obtained, as shown, using a Philips CM200-FEM microscope at several defoci. A through-magnification series of PPM images was used to locate the same fibre. From the TEM images the fibre width was determined to be 35 nm, which agrees well with its PPM shadow-image width. This magnification calibration can then be used to determine $z_1 = 600$ nm, $M = 150\,000$ for the PPM image in figure 7.4(a). The Fresnel number is therefore $F_N = \pi s^2/z_1\lambda = 34$ (s = width of fibre), compared with 154 ($\Delta f = -10\ \mu$m) for the TEM image in Figure 7.4(d).

For an opaque object, TEM and PPM images recorded at the same F_N should be identical, if the coherence conditions are similar, but the carbon fibre is a phase object for the electron beam in TEM, whereas it can be considered to be an opaque mask for the low energy electron beam in PPM. Thus, the object transmission functions in the two cases are entirely different.

The reconstruction of PPM images such as figure 7.4(a) using holographic techniques has been demonstrated elsewhere [40, 41, 44].

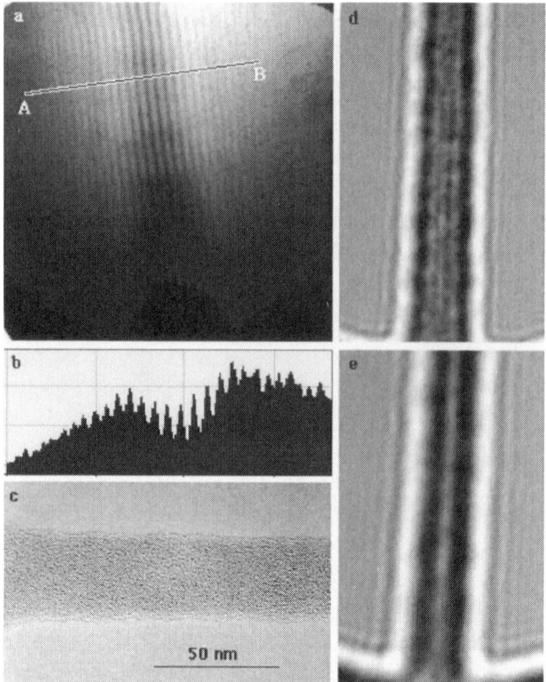

Figure 7.4   Fresnel fringes in PPM and TEM. (a) A point projection hologram formed by 40 eV electrons. Fibre width is 35 nm, Fresnel number $F_N = 34$. (b) Line profile from A to B. (c) In-focus TEM image of the same fibre using Philips CM200-FEG TEM. (d) TEM image of the fibre with defocus -10 μm. $F_N = 154$. (e) TEM image of the fibre with defocus -20 μm. $F_N = 77$.

7.5   DISCUSSION

The inflexibility of the PPM (in which magnification, voltage, defocus, field of view and beam current are not independent), the effects of charging, the limited number of pixels at low voltage, the problem of damage during focusing and sample preparation difficulties are the main challenges for the future development of the instrument. We discuss these in turn.

1. *Charging.* Localized surface and defect electron traps cause charging in insulating samples, and a variety of techniques have been described in the spectroscopy literature to deal with this (electron flood guns, doping, heating, ion beams, choice of voltage to balance currents etc). The most successful method for LEED and STM has been to work with very thin

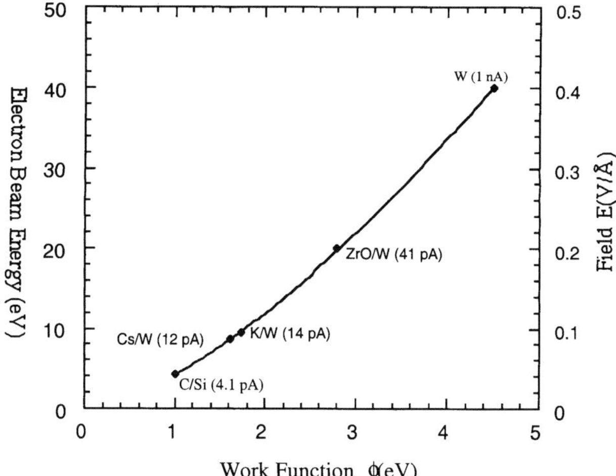

Figure 7.5 Electric field–work function plot. Extrapolated performance of tips identical to ours but with different work functions. The Fowler–Nordheim expression has been fitted to our experimental measurements at one point. Every point assumes the same tip shape and tip-to-sample distance. See also table 7.1.

films on a conducting substrate. For the PPM, charging can be minimized by studying conducting samples (such as the phthalocyanine semiconductors or tryptophan) or through the use of films much thinner than λ_i. In LEED, charging effects in thicker films commence only as the voltage is decreased below about 20 eV for ice, ammonia, naphthalene, benzene and n-paraffins. No charging is found at any thickness for phthalocyanines or tryptophan [31].

2. *Number of pixels.* As z_1 decreases the region of sample illuminated by the beam approaches the STM limit, and field of view is lost. One solution to this problem would be to operate the instrument as a scanning transmission electron microscope (STEM). Alternatively, by reducing the tip work function it would be possible to work at picoamp currents (as in recent LEED work [45]), rather than nanoamps, while retaining $z_1 > 100$ nm and a large field of view. Figure 7.5 shows the combination of currents and voltages, based on the Fowler–Nordheim expression, which would be possible using tips of different materials. These tips are assumed identical in sharpness to the experimental tungsten tip to which the results have been scaled.

It can be seen that a caesium-doped W tip would allow operation below the ESD threshold in the quasi-elastic region. To detect these small currents, a two-stage chevron channel plate (with gain 10^6) is proposed. The formation of a caesium-coated nanotip has been described [46]. These workers found both a reduction in energy spread and a reduction by one-quarter in voltage

Table 7.1 Datasheet for electric field–work function plot. $z_1 = 600$ nm. $E_F = 5$ eV in all the estimates for current.

Material	ϕ (eV)	E (V Å^{-1})	Beam (eV)	I (pA)
W	4.5	0.40	40	1000
ZrO/W	2.8	0.20	20	41
K/W	1.7	0.095	9.5	14
Cs/W	1.6	0.086	8.6	12
C/Si	1.0	0.042	4.2	4.2

needed for field emission, exactly in accord with the above extrapolation from our experimental measurements. A final solution is the use of lithographic techniques (such as the focused ion beam or FIB instruments) to make tips of extreme sharpness. By reducing the shank angle, the field at the field-emitting cluster of atoms can be increased, leading to higher fields at lower voltage [47].

3. *Focusing.* We have attempted unsuccessfully to avoid damage during the initial low magnification (high voltage) imaging by starting the instrument in STM mode on a nearby region, then drawing the tip back as voltage is increased. However, it is difficult (but not impossible) to obtain stable STM images of self-supporting holey carbon fibres. The AFM mode might also be used. A better approach is outlined in the next section.

4. *Sample preparation.* We plan to replace the holey carbon supports by readily available 60 nm thick SiN membranes across a hole in 3 mm diameter silicon discs. Smaller holes, about 10 nm in diameter, can be drilled in these membranes by FIB instruments, after carbon coating. The organic film will be laid over this array of small holes, and the PPM started in STM mode on the coated SiN. The longer inelastic mean free path in the quasi-elastic region greatly eases demands on sample preparation, so that films of up to 10 nm thickness, lying over very small holes, might be transparent. A second approach under investigation is based on the techniques of matrix isolation spectroscopy [13]. The idea is to embed biomolecules in thin films of solid xenon at 55 K, whose weak van der Waals' bonds leave the molecular shape unchanged. For xenon, $E_g = 9.3$ eV and $\phi_A = 0.5$ eV, so thin films will scatter almost entirely elastically at energies below 8.5 eV. This elastic scattering may also be eliminated (at some cost in resolution) by using an electron wavelength which is longer than the primitive lattice dimension, so that Bragg scattering is not possible within the xenon—the Ewald sphere diameter is smaller than the first-order reciprocal lattice spacing. Xenon films might be condensed on SiN films, to grow across holes. Our calculations for self-supporting long molecules across holes suggest that the inhomogeneity

of the electric field becomes a consideration for holographic reconstruction at high magnification [48]. The holographic reference wave can no longer be modelled as a spherical wave in the reconstruction. Alternatively, lipids that proteins bind to might be added to the DPPC—for example streptavidin, which binds to biotinylated lipids with high affinity and may be crystallized. Samples could be freeze dried.

One final possibility for damage-free imaging is suggested by the PPM instrument. It has frequently been pointed out that an electron beam pulse whose duration is shorter than the vibrational period of atoms in a sample (10^{-13} s) cannot couple energy to the lattice efficiently. Repetitive fast pulsing can build up an image. The small capacitance of the nanotip emitter may make this possible, using a fast laser to flood an SiO_2 photoconductor as a switch. Picosecond photoemmission electron microscopy has been achieved.

For molecules which can be crystallized, diffraction methods might also be applied at low voltage, below the damage threshold, in order to minimize damage. Multiple scattering and the phase problem then become important aspects of the data analysis. Since data are required over a range of energies to 'solve' structures by LEED, it is not practical to collect LEED data entirely in the quasi-elastic region. In addition, at these low energies, few spots are observed due to the very small Ewald sphere, and charging problems and stray magnetic fields present additional difficulties. The molecules may contain many more atoms than the $Si(7 \times 7)(111)$ structure, which is at the limit of present LEED capabilities. As an alternative, it has recently become possible to obtain convergent-beam LEED patterns on LEEM instruments [49]. These patterns, if obtained below the ESD threshold, might form the basis of a structure analysis method if enough intensity variation can be observed within the discs. This depends on the ratio $R = \xi_g^{-1} g^{-1}/(2\Theta_B)$ of the angular width of the two-beam rocking curve to twice the Bragg angle. Here ξ_g is the extinction distance for the first-order reflection g. Convergent beam electron diffraction (CBED) discs overlap if $R > 0.5$. Since $R = 2\gamma m_0 e V_g/(g^2 h^2)$ and so is proportional to the relativistic factor γ, it varies little between 10 eV and 100 kV. At 10 eV, intensity oscillations occur within an angular range equal to the Bragg angle for crystals with unit cells smaller than about 1 nm.

We reconcile these experimental results with our earlier observations on purple membrane (PM) at 100 V (in which no apparent damage was observed) as follows. Typical inelastic mean free path values for organics are about 2 nm at 100 eV. The thickness of uncoated, unstained PM is 5.2 nm, while that of our lipid is 2 nm. This suggests that, in the PM, the beam was entirely stopped (and material damaged) within a surface layer (2 nm thick), while the remainder of the structure remained intact, supporting the film, apparently undamaged. For the thinner lipid film, whose thickness is comparable to the inelastic mean free path, damage occurs more uniformly throughout the film, leading to its collapse.

7.6   CONCLUSIONS

1. Experimental data in the literature suggest that electron imaging of individual biomolecules may be possible in principle at an energy below the molecular dissociation threshold (about 7 eV for most hydrocarbons). These images will be limited in resolution by the electron wavelength ($\lambda = 0.46$ nm), but would be useful for determining the shape of individual molecules.
2. Unlike the conditions which normally apply in TEM, elastic electron scattering becomes much more probable than inelastic for hydrocarbons at low energies ($E_k < 20$ eV), especially in the quasi-elastic region.
3. Images of C16 lipids recorded by PPM at 110 V are almost opaque, and are destroyed by a dose of approximately 60 e$^-$ Å^{-2}. This is between 20 and 60 times greater than the dose which would destroy the films using 100 keV electrons.
4. Using picoamp currents and caesium-coated tungsten nanotips, it should be possible to obtain point projection images with a usefully large field of view, below the threshold energy for radiation damage. (Such an image of tobacco mosaic virus has very recently been obtained [50].)

ACKNOWLEDGMENT

This work was supported by ARO award DAAH04-96-1-0231.

REFERENCES

[1]   Crewe A, Wall J and Langmore J 1970 Observation of individual atoms by STEM *Science* **168** 1338–42

[2]   Breedlove J and Trammel G 1970 Molecular microscopy: fundamental limitations *Science* **170** 1310–12

[3]   Glaeser R M 1985 Electron crystallography of biological macromolecules *Ann. Rev. Phys. Chem.* **36** 243–301

[4]   Henderson R 1995 The potential and limitations of neutrons, electrons and X-rays for atomic resolution microscopy of unstained biological molecules *Q. Rev. Biophys.* **28** 171–99

[5]   Fujiyoshi Y and Uyeda N 1981 Direct imaging of double stranded DNA molecule *Ultramicroscopy* **7** 189–93

[6]   Shao Z 1996 Imaging organic molecules by AFM *Adv. Phys.* **45** 1–28

[7]   Cartier E, Pfluger P, Pireaux J and Vilar M 1987 Mean free paths and scattering for 0.1–4500 eV electrons in saturated hydrocarbon films *Appl. Phys.* A **44** 43–52

[8] Rowntree P, Parenteau L and Sanche L 1991 Anion yields produced by low energy electron impact on condensed hydrocarbon films *J. Phys. Chem.* **95** 4902–9

[9] Leclerc G, Cui Z and Sanche L 1987 Effective dissociation cross section for low energy (0.5–31 eV) electron impact on solid *n*-hexane thin films *J. Chem. Phys.* **91** 6461–9

[10] Howie A, Mohd Muhid M N, Rocca F J and Valdre U 1987 Beam damage in organic crystals *EMAG'87 Inst. Phys. Conf. Ser.* vol 90 (Bristol: Institute of Physics) pp 155–7

[11] Seah M and Dench W 1979 Inelastic electron path lengths *Surf. Interface Anal.* **1** 2–25

[12] Binh V T 1995 Nanometric observations at low energy by Fresnel projection microscopy. Carbon and polymer fibres *Ultramicroscopy* **58** 307–13

[13] Sanche L, Perluzzo G and Michaud M 1985 Electron transmission spectroscopy of matrix isolated N_2 *J. Chem. Phys.* **83** 3837–48

[14] Spence J C H 1997 STEM and shadow imaging of biomolecules at 6 eV beam energy *Micron* **28** 101–16

[15] Bader G, Perluzzo G, Caron L and Sanche L 1982 Elastic and inelastic mean free path determination in solid xenon from electron transmission experiments *Phys. Rev.* **26** 6019–23

[16] Bader G, Perluzzo G, Caron L and Sanche L 1984 Structural order effects in low energy transmission spectra of condensed Ar, Kr, N_2, Co and O_2 *Phys. Rev.* B **30** 78–84

[17] Sommer A H 1968 *Photoemissive Materials* (New York: Wiley)

[18] Gullikson E M and Henke B L 1989 X-ray induced secondary emission from solid xenon *Phys. Rev.* B **39** 1–6

[19] Howie A 1995 Recent developments in secondary electron imaging *J. Microsc.* **180** 192–9

[20] Perluzzo G, Sanche L, Gaubert C and Baudoing R 1984 Thickness-dependent interference structure in the 0–15 eV electron transmission spectra of rare-gas films *Phys. Rev.* B **30** 4292–9

[21] Anderson S 1969 Band gaps in LEED *Surf. Sci.* **18** 325–9

[22] Michaud M, Sanche L, Gaubert C and Baudoing R 1988 Low energy electron reflection and transmission on Ar films *Surf. Sci.* **205** 447–59

[23] Qian W, Spence J and Zuo J M 1993 Theory of transmission low energy diffraction *Acta Crystallogr.* A **49** 436–45

[24] Tanuma S, Powell C J and Penn D R 1991 Calculations of electron inelastic mean free paths *Surf. Interface Anal.* **17** 911–9
(Also, see Meepagala S and Baykul M 1994 Ballistic electron emission into vacuum from an STM tip through free-standing gold films *Phys. Rev.* B **50** 13 786–94 for a demonstration of transmission of low voltage electrons through a thin metal film)

[25] Kwei C, Chen Y, Tung C and Wang J 1993 Electron inelastic mean free paths for plasmon and interband transitions *Surf. Sci.* **293** 202–9

[26] Sparks M, Mills D, Warren R, Holstein T, Maradudin A, Sham L, Loh E and King D 1981 The electron–phonon interaction in crystals *Phys. Rev. B* **24** 3519–23

[27] Frohlich H 1954 Theory of the dielectric function *Adv. Phys.* **3** 325–32

[28] Isaacson M, Johnson D and Crewe A V 1973 Electron beam excitation and damage of biological molecules *Radiat. Res.* **55** 205–15

[29] Howie A, Rucca F J and Valdre U 1985 Electron beam ionization damage processes in p-terphenyl *Phil. Mag.* **52** 751–66

[30] Christophrou L 1971 *Atomic and Molecular Radiation Physics* (New York: Wiley)

[31] Firment L and Somorjai G 1979 Low energy electron diffraction studies of the surfaces of molecular crystals *Surf. Sci.* **84** 275–83

[32] Goulet T, Pou V and Jay-Gerin J 1986 A procedure for determining low energy (1–10 eV) mfp in molecular solids—benzene *J. Electron Spectrosc. Relat. Phenom.* **41** 157–66

[33] Spence J C H 1988 *Experimental High Resolution Electron Microscopy* (New York: Oxford University Press)

[34] Goulet T and Jay-Gerin J 1985 Low energy (1–10 eV) electron transmission through thin solid benzene films *Solid State Commun.* **55** 619–723

[35] Zimmerman U and Karl N 1992 Epitaxial growth of coronene: a LEED study *Surf. Sci.* **268** 296–309

[36] Fukami A and Adachi K 1965 A new method of preparation of a self-perforated micro plastic grid and its application *J. Electrom Microsc.* **14** 112–8

[37] Spence J C H, Qian W and Melmed A 1993 Low voltage point projection microscopy and its possibilities *Ultramicroscopy* **52** 473–489

[38] Spence J 1996 Low voltage point projection microscopy *Handbook of Microscopy* ed S Amelincx and D Van Dyke (New York: Elsevier)

[39] Spence J C H 1992 Convergent-beam nano-diffraction, in-line holography and coherent shadow imaging *Optik* **92** 57–63

[40] Spence J, Zhang X and Qian W 1995 On the reconstruction of low voltage point projection holograms *Electron Holography* ed A Tonomura *et al* (New York: Elsevier) pp 267–74.

[41] Bleloch A L, Howie A and James E M 1997 Application of the Gerchberg–Saxton algorithm to low voltage electron holograms *Appl. Surf. Sci. Proc. IVESC Conf.* **111** 180–4

[42] Paganin D and Nugent K 1998 Noninterferometric phase imaging with partially coherent light *Phys. Rev. Lett.* **80** 2586–94

[43] Morin R and Gargani A 1993 Ultra low energy electron projection holograms *Phys. Rev. B* **48** 6643–5

[44] Kreuzer H J, Nakamura K, Wierzbicki A, Fink H-J and Schmid H 1992

Theory of the point source electron microscope *Ultramicroscopy* **45** 381–98

[45] Ogletree D, Blackman G, Hwang R, Starke U, Somorjai G and Katz J 1992 A new pulse counting LEED system *Rev. Sci. Instrum.* **63** 104–17

[46] Morin R and Fink H 1994 Highly monochromatic electron point-source beams *Appl. Phys. Lett.* **65** 2362–74

[47] Scheinfein M, Qian W and J C H Spence 1993 Aberrations of emission cathodes: nanometer diameter field emission electron sources *J. Appl. Phys.* **73** 2057–68

[48] Weierstall U, Huang X and Spence J 1997 Twin image supression and field inhomogeneity in in-line electron holography *Proc. EMSA 1997* ed G Bailey (New York: Springer) pp 1143–4

[49] Tromp R 1997 Personal communication

[50] Weierstall U, Spence J, Stevens M and Downing K 1999 Point-projection imaging of TMV at 40 eV electron energy *Micron* at press

8

DIFFRACTION IMAGING USING BACKSCATTERED ELECTRONS: FUNDAMENTALS AND APPLICATIONS

S L Dudarev

8.1   INTRODUCTION

The problem of backscattering of high energy electrons from a crystalline material has been attracting the attention of researchers working in the field of electron microscopy and materials science for more than 30 years. The complexity of its physical content complemented with the important part played by electron backscattering in a large number of applications in materials science makes it a very interesting problem to study. To model the process of electron backscattering it is necessary to develop accurate mathematical methods suitable for the description of multiple scattering of electrons in a crystal. At the same time electron backscattering can be viewed as an efficient practical tool for obtaining information about structural properties of surface layers of crystalline materials. The long history of the development of experimental techniques employing electron backscattering as well as corresponding theoretical methods has witnessed many important advances, and some of them have taken place only recently.

What makes the problem of electron backscattering so complex and interesting? On the one hand, it is relatively easy to obtain a beam of electrons accelerated to the energy of a few tens of keV and to focus it on a surface, and this represents a real advantage of the method. On the other hand, electrons are light particles and therefore in a solid their trajectories do not look like straight lines (provided that we may talk about electron trajectories in a crystal where it is essential to consider quantum-mechanical diffraction effects), and this represents a serious drawback since it shows that simple theoretical models may often be unsuitable for interpreting the results of experimental observations.

In this chapter I would like to address mainly the problem of theoretical description of electron backscattering by a crystalline material, comparing, where possible, theoretical results with experimental data. In particular, I am going to illustrate the important part played by the computer hardware in the evolution of our understanding of electron backscattering. Many questions which remained beyond our reach 20 years ago can now be answered on the basis of a straightforward and relatively simple numerical simulation. In this respect, progress in the theoretical understanding of various aspects of electron backscattering appears to have matched the development in scanning electron microscope (SEM) imaging where a number of remarkable results have been obtained over the last decade.

In choosing examples for the discussion I followed the idea that to make a complex problem understandable it is necessary to explain to the reader the underlying basic principles first and then to illustrate them using some suitable examples. In the case of electron backscattering from a crystal the formulation of the basic principle (i.e. the equation describing the evolution of the distribution of incident electrons under the influence of incoherent scattering by phonons and dynamical diffraction by the average periodic potential of the crystal) requires the introduction of several new concepts, in the first place the concept of the density matrix of high energy electrons. The density matrix generalizes the ordinary probability distribution of electrons over momentum and coordinate to the case where it is necessary to introduce the notion of *coherence* and *coherent scattering* by periodic arrays of atoms.

I shall start by describing how to treat backscattering of electrons by an amorphous material. In this case practically all of the known experimental facts can be explained using the (classical) Boltzmann transport equation. To study electron backscattering by a crystalline material it is necessary to include diffraction effects in the transport equation. This can be achieved by generalizing the transport equation to the quantum-mechanical case and by introducing the quantum kinetic equation for the density matrix of high energy electrons. Once this equation is derived, the whole problem reduces to the development of mathematical tools for solving it. This includes using either purely numerical approaches or approximate analytical methods. In particular, below I am going to show how the quantum kinetic equation may be solved perturbatively in the case where we are interested in simulating scanning electron microscope electron channelling contrast images (ECCIs) of crystal defects.

Since the amount of reference material on electron backscattering is very extensive, it is hard to expect that every relevant publication can be referred to in a short article. I shall therefore focus attention mainly on more recent results and on those that I consider important for illustrating the main features of the phenomenon of electron backscattering. For further reading I suggest two good review articles published in the *Journal of Applied Physics* in 1982. One of them is written by Niedrig [1] and the other is by Joy *et al* [2]. The first deals with scattering of electrons by an amorphous material while the second review is entirely devoted to the description of diffraction effects in electron backscattering. A survey of relevant experimental techniques can be found in a series of review articles by Booker [3]. Dudarev [4] described a range of mathematical methods developed in the theory of scattering of waves by partially disordered periodic systems. An excellent account of recent developments in the field of electron channelling contrast imaging is given in a review by Wilkinson and Hirsch [5].

8.2   ELECTRON BACKSCATTERING FROM AN AMORPHOUS SOLID

We start by considering how to describe backscattering of electrons by an amorphous solid. Why do we need to think about this, at first glance not particularly relevant, problem? To answer this question it is necessary to appeal to the reader's physical imagination. What is the mechanism responsible for backscattering of high energy electrons? Can the experimentally observed distribution of electrons as a function of angles and energy result from backscattering by a periodic ensemble of atoms?

Here is a simple proof showing that the assumption that electrons are backscattered as a result of purely elastic interaction with a periodic ensemble of atoms does not agree with the available experimental information. Indeed, since the potential of any periodic array of atoms may be expanded in a two-dimensional Fourier series in the plane parallel to the surface (in other words, any crystal potential can be represented by a *two-dimensionally* periodic function in a plane parallel to the plane of the surface) it is clear that in any process of purely elastic interaction of an incident electron with the crystal the projection of the momentum parallel to the surface is going to be either conserved or it is going to be changed by $\hbar g$, where g is a reciprocal lattice vector involved in the above two-dimensional Fourier expansion. The term 'elastic interaction' implies that the total energy is conserved and therefore the momentum of the electron backscattered from the crystal must lie on the surface of a sphere, the radius of which equals the length of the vector representing the momentum of the incident electrons (see figure 8.1). This analysis shows that in the case where the interaction of incident electrons with the crystal is purely elastic, the only possibility is that the distribution of backscattered electrons has the form of a countable set of beams reflected from the surface in the backward direction. On a fluorescent screen this distribution is going to be seen as a set of bright diffraction spots with zero background intensity between the spots.

The experimentally observed distribution of electrons backscattered from a crystal is entirely different. Diffraction spots are in fact never seen in experiments on electron backscattering (unless the energy of electrons is very low or unless we study diffraction at very low glancing angles of the order of one or two

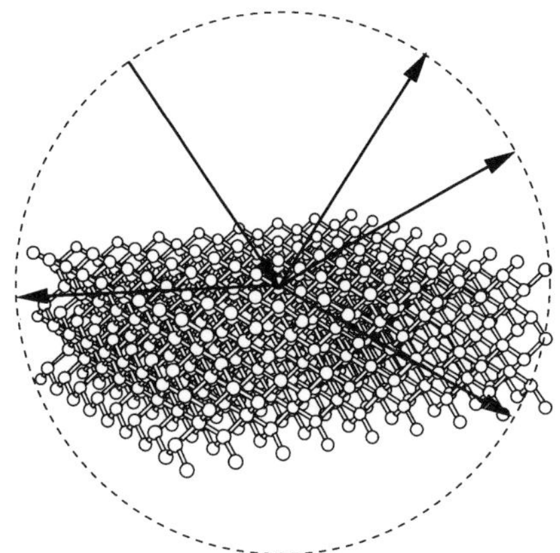

Figure 8.1 Schematic diagram illustrating the hypothetical process of purely elastic (i.e. diffraction) backscattering of electrons from a crystal. In the case shown in this figure the projection of the momentum transfer on a plane parallel to the surface of the crystal equals $\hbar g$, where g is a two-dimensional reciprocal lattice vector.

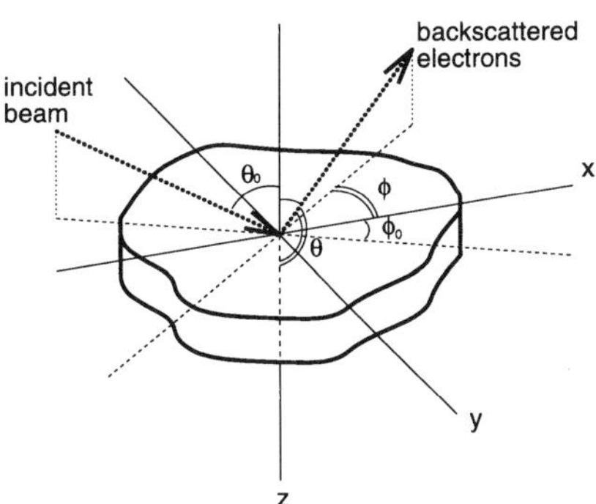

Figure 8.2 Schematic diagram illustrating the geometry of electron backscattering.

degrees). The distribution of backscattered electrons over angle and energy observed experimentally has the form of a continuous diffuse background spanning over the entire backward hemisphere $0 \leqslant \phi \leqslant 2\pi$, $\pi/2 \leqslant \theta \leqslant \pi$ (see figure 8.2 which illustrates the geometry of scattering and the notation used). In other words, the distribution of electrons backscattered from a crystal looks similar to the distribution of electrons backscattered by a material where the presence of crystalline order can be neglected (this case is often referred to as the case of scattering by an amorphous material).

What are the physical assumptions involved in the treatment of electron backscattering by an amorphous material? The most important one consists in saying that since we want to describe experiments where we would look at backscattering of not just a few, but of very many electrons, then it is the concept of the probability distribution of electrons over the momentum and coordinate $f(p, r)$ which represents the central entity of the theory, see for example [6] and [7] for more detail. To find the density $n(R)$ of electrons at a given point R one needs to integrate $f(p, R)$ over the momentum p:

$$n(R) = \int \mathrm{d}p f(p, R). \tag{8.1}$$

Using $f(p, R)$ we can also define the density of the current flow of electrons

$$N(p, R) = \frac{p}{m} f(p, R) \tag{8.2}$$

where m is the electron mass. $N(p, R)$ represents the number of electrons with momentum p passing through a unit area in the direction normal to this area. It is often more convenient to consider the distribution of electrons over the direction of scattering $n = p/p$ and energy E. The corresponding function $N(n, E, r)$ is also defined by equation (8.2) as the distribution function $f(n, E, r)$ multiplied by the velocity of electrons (also expressed as a function of the energy of electrons).

Function $N(n, E, r)$ plays a central part in the theory of electron backscattering. The reason for this is that there exists a closed equation describing the evolution of $N(n, E, r)$. This equation, representing a linearized version of the Boltzmann equation, is called the transport equation

$$n \frac{\partial}{\partial r} N(n, E, r) = n_V \int \mathrm{d}o_{n'} \int \mathrm{d}\epsilon \frac{\mathrm{d}^2\sigma}{\mathrm{d}o_{n'}\mathrm{d}\epsilon}(n, E|n', E + \epsilon) N(n', E + \epsilon, r)$$

$$- n_V \int \mathrm{d}o_{n'} \int \mathrm{d}\epsilon \frac{\mathrm{d}^2\sigma}{\mathrm{d}o_{n'}\mathrm{d}\epsilon}(n', E - \epsilon|n, E) N(n, E, r). \tag{8.3}$$

This equation describes how the distribution of electrons over solid angle and energy evolves when electrons collide with atoms, n_V being the number of atoms per unit volume. In (8.3) n' and n are the unit vectors characterizing the direction of motion of an electron before and after a collision with an atom. $\mathrm{d}^2\sigma(n, E|n', E + \epsilon)/\mathrm{d}o_{n'}\mathrm{d}\epsilon$ is the double differential cross section of scattering of an electron by an atom, where ϵ is the energy transferred to the atom in a collision and $E + \epsilon$ and E are the energies of the electron before and after the inelastic interaction with an atom. The main assumption involved in the derivation of equation (8.3) consists in the supposition that the average distance travelled by an electron between successive collisions with atoms is many times the electron wavelength and the average size of an atom, see Abrikosov *et al* [8].

A number of approaches were developed in the past to solve equation (8.3) for the case of electron backscattering. Comprehensive studies of this problem based on equation (8.3) have been performed by Wang and Guth [9] and more recently by Tilinin [10] and by Tilinin and Werner [11, 12]. A good description of recent theoretical developments can be found in a review article by Tilinin and Werner [13]. Since the approach developed by the latter authors makes it possible to obtain an approximate analytical solution of the problem, I shall briefly describe it here.

I shall start by introducing an approximation which is commonly used in the theory of multiple scattering of charged particles in condensed matter. This approximation consists in representing the double differential cross section of scattering $\mathrm{d}^2\sigma/\mathrm{d}o_{n'}\mathrm{d}\epsilon$ in the form of a sum of two terms. The first term describes *purely elastic* scattering of an electron by an atom, while the second term represents the cross section of energy losses. It is assumed that the loss of energy is not associated with the change in the direction of motion of the electron interacting with an atom. Of course, such separation of the exact differential cross section of scattering into two parts is somewhat artificial and there are processes (like inelastic collisions of incident electrons with atomic electrons) that give rise to both energy losses *and* to the change of the direction in which the electron moves. However, in the problem of electron backscattering where *multiple* scattering plays the dominant part, the accuracy of the above approximation is in most cases sufficient. The simplification in the mathematical treatment which can be achieved as a result of adopting the approximate representation of the cross section far outweighs the error introduced by this approximation.

Consider the case of normal incidence of electrons on the surface of an amorphous material. In this geometry of scattering the solution of the transport equation depends only on the energy E of electrons, the coordinate z in the direction normal to the surface and the polar angle θ. By introducing a convenient variable $\mu = \cos\theta$, we can write the transport equation in the form

$$\mu\frac{\partial}{\partial z}N(\mu, E, z) = -n_V[\sigma_{\mathrm{el}}(E)+\sigma_{\mathrm{inel}}(E)]N(\mu, E, z)$$

$$+n_V\int \mathrm{d}o_{n'}\frac{\mathrm{d}\sigma_{\mathrm{el}}}{\mathrm{d}o_{n'}}(\boldsymbol{n}|\boldsymbol{n}')N(\mu', E, z)$$

$$+n_V\int_0^{\epsilon_{\max}} \mathrm{d}\epsilon\frac{\mathrm{d}\sigma_{\mathrm{inel}}}{\mathrm{d}\epsilon}(E|E+\epsilon)N(\mu, E+\epsilon, z) \tag{8.4}$$

where $\sigma_{\mathrm{el}}(E)$ and $\sigma_{\mathrm{inel}}(E)$ are the total cross sections of elastic and inelastic interaction, respectively, and $\epsilon_{\max}$ is the maximum possible energy loss occurring in a single collision of the electron with an atom. The boundary condition for equation (8.4) is

$$N(\mu, E, z = 0) = \begin{cases} N_0 \delta(1 - \mu)\delta(E - E_0) & \text{when } \mu > 0 \\ N_0 S(\mu, E) & \text{when } \mu < 0 \end{cases} \qquad (8.5)$$

where N_0 is the current density associated with the incident beam of electrons, E_0 is the energy of the incident electrons and $S(\mu, E)$ is an *unknown* function describing the distribution of backscattered electrons as a function of the angle of scattering and energy.

To find this distribution we need to solve equation (8.4) supplemented with boundary condition (8.5). Below we describe how this can be accomplished using a numerical approach. Here we show how an approximate analytical solution of the problem can be found using the so-called 'transport' approximation proposed by Tilinin in 1982 [10]. To understand the range of validity of this approximation consider the evolution of the distribution of electrons incident on the surface of a slab of an amorphous material. As electrons enter the solid they continue moving along a straight line. However, gradually their distribution becomes more and more isotropic and at a certain depth which approximately equals the *transport* mean free path $l_{tr} = (n_V \sigma_{tr})^{-1}$, where σ_{tr} is the transport cross section of scattering (see §139 of Landau and Lifshits [14]), the distribution of electrons over the angle of scattering becomes nearly isotropic. We take this into account and instead of treating effects of elastic scattering exactly as in equation (8.4), we approximate the relevant term in the right-hand side of (8.4) assuming that the *anisotropy* of the function $N(\mu, E, z)$ is small. This leads to the 'transport' approximation according to which equation (8.4) is written as

$$\mu \frac{\partial}{\partial z} N(\mu, E, z) = -n_V [\sigma_{tr}(E_0) + \sigma_{inel}(E_0)] N(\mu, E, z)$$

$$+ \frac{1}{2} n_V \sigma_{tr}(E_0) \int_{-1}^{1} N(\mu', E, z)$$

$$+ n_V \int_{0}^{\epsilon_{max}} d\epsilon \frac{d\sigma_{inel}}{d\epsilon} (E_0 | E_0 + \epsilon) N(\mu, E + \epsilon, z) \qquad (8.6)$$

where σ_{tr} is the transport cross section and where its dependence on energy is neglected. Since equation (8.6) formally coincides with the equation describing multiple scattering of particles by pointlike randomly distributed centres, its solution can be readily found by applying methods developed by Chandrasekhar in the 1940s [15] (in the present case it is also necessary to evaluate the Laplace transform of the solution over the energy variable in order to take energy losses into account).

Having solved (8.6) we find that the distribution $S(\mu, E)$ of electrons backscattered from an amorphous material is nearly isotropic (in other words, the cross section of backscattering $|\mu|S(\mu, E)$ follows the cosine law). This agrees well with experimental observations and justifies the validity of the approximations made above. The transport approximation also makes it possible to find a simple formula for the total coefficient of electron backscattering

$$R_{\text{tot}} = \int\limits_{-1}^{0} \mathrm{d}\mu\, |\mu| \int \mathrm{d}E\, S(\mu, E). \tag{8.7}$$

The total coefficient of electron backscattering (i.e. the probability that an incident electron is not going to lose its entire energy inside the crystal and that it will eventually emerge back into the vacuum) evaluated for the case of normal incidence equals

$$R_{\text{tot}} = 1 - \frac{1}{\sqrt{1+s}} H\left(1, \frac{s}{s+1}\right)$$

$$s = (Z/8)L_{\text{C}}(E_0)/L_{\text{ion}}(E_0) \tag{8.8}$$

$$H(1, \lambda) = \frac{1 + \sqrt{3}}{1 + \sqrt{3(1 - \lambda)}}[1 + 0.03\lambda(1 + \lambda^3)]$$

where Z is the average atomic number of the material and $L_{\text{C}}(E_0)$ and $L_{\text{ion}}(E_0)$ are the Coulomb and ionization logarithms, respectively. Values calculated using this equation and the ratio $L_{\text{C}}(E_0)/L_{\text{ion}}(E_0)$ given by Tilinin and Werner [13] agree well with the values evaluated numerically by solving the Boltzmann equation or obtained by carrying out Monte Carlo simulations of electron backscattering. The agreement between the analytical treatment of the problem, numerical studies, and experimental data shows that the physical assumptions involved in the mathematical treatment of electron backscattering are sound and that we are now in the position to address the question of how they can be generalized to the case of electron backscattering by a crystalline material.

8.3 BACKSCATTERING FROM A CRYSTAL: THE ROLE OF PHONONS

The first question which needs to be addressed within the framework of the treatment of electron backscattering by a crystalline material concerns the origin of the effect itself. We have already shown above that diffraction alone cannot explain the formation of the nearly isotropic distribution of backscattered electrons over the angle of scattering observed experimentally (see e.g. a review by Joy *et al* [2]). It is therefore scattering by *fluctuations* of the crystal potential rather than by the average potential itself which gives rise to electron backscattering from a crystal. Experimental studies performed by Boersch *et al* [16, 17] and by Makarov *et al* [18] have confirmed the presence of the *temperature-dependent* broadening of the energy spectrum of backscattered electrons. The dependence of the width of the quasielastic peak in the energy spectrum on the angle of scattering and on the absolute temperature shows that electrons are backscattered by *thermal fluctuations* of the interaction potential and that the phenomenon of electron backscattering is associated with the creation of phonon excitations in impact collisions between incident electrons and atoms in a crystal.

To illustrate this statement we consider a collision between an electron and an atom localized at a lattice site in a crystal. The differential cross section of scattering is given by (here we disregard excitations of the electronic degrees of freedom of the atom and assume that the change in the energy of the backscattered electron is entirely due to the change in the vibrational state of the nucleus)

$$\frac{\mathrm{d}^2\sigma}{\mathrm{d}o\mathrm{d}E} = \frac{1}{\hbar}\left(\frac{\mathrm{d}\sigma}{\mathrm{d}o}\right) S_{\mathrm{ph}}\left(\frac{\boldsymbol{p}_1 - \boldsymbol{p}_2}{\hbar}, \frac{\Delta E}{\hbar}\right) \tag{8.9}$$

where $\boldsymbol{p}_1$ and $\boldsymbol{p}_2$ are the initial and final momenta of the fast electron and $S_{\mathrm{ph}}(\boldsymbol{q}, \omega)$ is the momentum- and frequency-dependent phonon dynamic structure factor

$$S_{\mathrm{ph}}(\boldsymbol{q}, \omega) = \sum_{l,n} \frac{1}{Z} \exp(-E_l/k_B T)|\langle l| \exp(-i\boldsymbol{q} \cdot \boldsymbol{u})|n\rangle|^2 \delta\left(\omega + \frac{E_l - E_n}{\hbar}\right). \tag{8.10}$$

In (8.10) vector $\boldsymbol{u}$ denotes the thermal displacement of the atom, $\boldsymbol{q} = (\boldsymbol{p}_1 - \boldsymbol{p}_2)/\hbar$, $Z = \sum_l \exp(-E_l/k_B T)$ and summation over l and n is performed over the states of the phonon subsystem of the crystal. For a perfect crystal the dynamic structure factor can be evaluated using a method developed in the theory of the Mössbauer effect and described by Kagan [19]

$$S_{\mathrm{ph}}(\boldsymbol{q}, \omega) = \frac{1}{2\pi} \int_{-\infty}^{\infty} \mathrm{d}t \, \exp[i\omega t + \phi(t) - \phi(0)] \tag{8.11}$$

where

$$\phi(t) = \frac{R}{N_V} \sum_{\boldsymbol{f},\alpha} \frac{|\boldsymbol{v}_q \cdot \boldsymbol{e}(\boldsymbol{f}, \alpha)|^2}{\hbar\omega(\boldsymbol{f}, \alpha)} \tag{8.12}$$

$$\times \{[\bar{n}(\boldsymbol{f}, \alpha) + 1]\exp[-i\omega(\boldsymbol{f}, \alpha)t] + \bar{n}(\boldsymbol{f}, \alpha)\exp[i\omega(\boldsymbol{f}, \alpha)t]\}.$$

In this formula $R = \hbar^2 q^2/2M$ is the recoil energy, M is the mass of the atom, $\boldsymbol{v}_q = \boldsymbol{q}/q$ is the unit vector in the direction of $\boldsymbol{q}$, N_V is the number of crystal unit cells per unit volume and $\bar{n}(\boldsymbol{f}, \alpha) = [\exp(\hbar\omega(\boldsymbol{f}, \alpha)/k_B T) - 1]^{-1}$

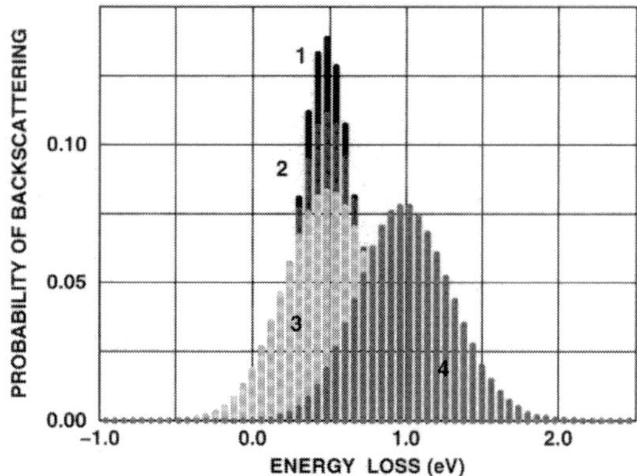

Figure 8.3 Phonon recoil energy loss spectra of electrons backscattered from a crystal calculated using equation (8.13). Parameters used in the calculation are: (1) $R = 0.5$ eV, $T = 80$ K, $\hbar\omega_0 = 0.06$ eV; (2) $R = 0.5$ eV, $T = 300$ K, $\hbar\omega_0 = 0.06$ eV; (3) $R = 0.5$ eV, $T = 600$ K, $\hbar\omega_0 = 0.06$ eV; (4) $R = 1.0$ eV, $T = 300$ K, $\hbar\omega_0 = 0.06$ eV. Note the effect of *multiple* generation of phonons in a single event of scattering of the incident electron by an atom in a crystal.

is the average occupation number of phonon states. Since the magnitude of q is defined by the geometry of scattering, we concentrate on the dependence of (8.9) on energy $\hbar\omega$. If we disregard the dispersion of phonon frequencies and assume $\omega(f, \alpha) = \omega_0 = \text{const}$, we obtain

$$S_{\text{ph}}(q, \omega) = \sum_{L=-\infty}^{\infty} \delta(\omega - \omega_0 L) \exp\left[-\frac{R}{\hbar\omega_0} \coth\left(\frac{\hbar\omega_0}{2k_B T}\right)\right] \tag{8.13}$$

$$\times \exp\left(\frac{\hbar\omega_0}{2k_B T} L\right) I_L \left\{2\frac{R}{\hbar\omega_0} \exp\left(\frac{\hbar\omega_0}{2k_B T}\right)\left[\exp\left(\frac{\hbar\omega_0}{k_B T}\right) - 1\right]^{-1}\right\}$$

where $I_L(x)$ is the modified Bessel function and $L = 0, \pm 1, \pm 2, \ldots$. A schematic sketch of the spectrum described by equation (8.13) is shown in figure 8.3. The dependence of the half-width of the spectrum (8.13) on the temperature is given by

$$\langle(\hbar\omega - R)^2\rangle = R\hbar\omega_0 \coth\left(\frac{\hbar\omega_0}{2k_B T}\right). \tag{8.14}$$

A more accurate expression for the half-width of the spectrum which takes into account the dispersion of phonon frequencies can be obtained directly from (8.11) via

$$\langle\omega^n\rangle = \int_{-\infty}^{\infty} d\omega\, \omega^n S_{\text{ph}}(q, \omega) = \left(i\frac{\partial}{\partial t}\right)^n [\exp(\phi(t) - \phi(0))]|_{t=0} \tag{8.15}$$

which gives

$$\hbar\langle\omega\rangle = R$$

$$\langle(\hbar\omega - R)^2\rangle = \frac{R}{N_V}\sum_{f,\alpha}|v_q \cdot e(f,\alpha)|^2\hbar\omega(f,\alpha)[2\bar{n}(f,\alpha) + 1]. \tag{8.16}$$

Values given by (8.16) and (8.14) converge in the classical limit $k_B T \gg \hbar\omega_0$

$$\langle(\hbar\omega - R)^2\rangle = 2k_B T R \tag{8.17}$$

which agrees with the dependence observed experimentally by Boersch [17], by Makarov *et al* [18] and by Igonin and Makarov [20]. This confirms that phonon excitations are indeed responsible for the formation of the flux of electrons backscattered from a crystalline material.

8.4   THE FORMATION OF ELECTRON CHANNELLING PATTERNS

At present we know that (i) the distribution of electrons backscattered by a crystal at room temperature looks similar to the distribution of electrons backscattered by an amorphous solid, (ii) phonon excitations are responsible for the formation of the flux of backscattered electrons and (iii) the difference between the rates of electron energy losses in an amorphous material and in a crystalline solid is negligibly small. Given these three points, we shall now attempt to construct an equation which we hope will be suitable for the description of incoherent scattering of high energy electrons by phonons, and will also take into account energy losses *and* diffraction of electrons by the periodic distribution of atoms in a crystal. It is clear that the derivation of this equation must be based on the Schrödinger equation (since diffraction effects are entirely absent in the Boltzmann equation (8.3) and we know that the treatment of these effects must rely upon a consistent quantum-mechanical approach). At the same time it is clear that in the limiting case of scattering of electrons by a random distribution of atoms the solution of this new equation must coincide with that of the Boltzmann equation obtained for the same geometry of scattering. A similar conceptual problem is known in the theory of radiative transfer, see Frisch [21], where the question about the relationship between the wave optics and the transport of radiation in a turbid medium represents one of the fundamental issues. The method of deriving the transport equation from the wave equation discovered in the theory of wave propagation in a random medium shows that the main entity involved in the formulation of the *quantum-mechanical* transport theory is the *density matrix* (sometimes also called the *mutual coherence function*) $\rho(r, r') = \langle\psi(r)\psi^*(r')\rangle$, where $\psi(r)$ is the wave function and the brackets $\langle ...\rangle$ denote averaging over the distribution of atoms in the scattering medium. The meaning of this new quantity $\rho(r, r')$ is relatively simple: its diagonal elements $\rho(r, r)$ equal the average probability of finding an electron at a point r while its off-diagonal elements $\rho(r, r')$ for $r \neq r'$ represent a measure of *coherence* of the wave field of electrons at these two points. In most cases the off-diagonal elements of $\rho(r, r')$ vanish rapidly as the distance between r and r' increases, but in certain cases (a famous one is described in a book by Mahan [22]) the off-diagonal elements of the density matrix remain finite even if the two points r and r' are separated by an infinitely large distance. For a plane wave $\exp(i k \cdot r)$ the density matrix has the form $\rho(r, r') = \exp[i k \cdot (r - r')]$ and the modulus of its off-diagonal elements $|\rho(r, r')|$ does not depend on the distance between r and r'. It can be shown that if the crystal potential contains no random component, the density matrix retains this property and the wave field of electrons remains entirely coherent everywhere in real space. However, scattering of electrons by statistical fluctuations of the potential destroys the coherence of the wave field of electrons and hence the off-diagonal elements of the density matrix become functions vanishing in the limit $|r - r'| \to \infty$.

To establish a link between wave optics and the transport theory it is often assumed that $\rho(r, r')$ can be represented in the functional form

$$\rho(r, r') = \Phi[r - r'; (r + r')/2] \tag{8.18}$$

where the rate of variation of Φ as a function of its first variable is many times the rate of variation of this function with respect to its second variable. Given this assumption, the Fourier transform of $\Phi[r - r'; (r + r')/2]$ with respect to $r - r'$ leads to an equation which resembles the Boltzmann equation (see Lifshits and Pitaevskii [24] for more detail). In some cases this procedure gives rise to a reasonably accurate description of scattering of waves by a disordered ensemble of scattering centres. At the same time it is clear that it is hard to think of a consistent approach to setting boundary conditions for this equation, see Gorodnichev *et al* [23]. As a result, in the problem of electron backscattering from a crystalline material it appears to be necessary to solve the equation for the density matrix $\rho(r, r', E)$ as it stands, making no linear transformation of the two coordinates r and r'. A derivation of the quantum kinetic equation for the density matrix of high energy electrons $\rho(r, r', E)$ describing diffraction and multiple incoherent scattering in a crystalline material has been given by Dudarev, Peng and Whelan [25]. This equation has the form

$$\rho(r, r', E) = \rho_0(r, r', E) + \iint d^3x\, d^3x'\, G(r, x, E)\, G^*(r', x', E)$$

$$\times \left[\int d\omega\, \bar{s}(x', x, \omega)\, \rho(x, x', E + \hbar\omega) \right]. \tag{8.19}$$

Here $\rho_0(r, r', E)$ is the one-particle density matrix of the incident electrons (in some cases it can be represented simply in the form of a product of two wave functions multiplied by the delta function $\delta(E - E_0)$). $G(r, x, E)$ is the *probability amplitude* of propagation of an electron between the two points x and r in the crystal. In equation (8.19) $\bar{s}(x', x, \omega)$ is the mixed dynamic form factor describing the interaction of the incident electron with phonon and electronic excitations.

If we assume that thermal vibrations of different atoms are independent and neglect the recoil energy, the mixed dynamic form factor can be written in the form

$$\bar{s}(x', x, \omega) = \delta(\omega) \sum_a \int\int \frac{\mathrm{d}^3q\,\mathrm{d}^3q'}{(2\pi)^6} U_a(q)U_a(-q')\exp[iq\cdot(x-r_a)]$$

$$\times\exp[iq'\cdot(x'-r_a)]\left\{\exp[-M_a(q-q')] - \exp[-M_a(q) - M_a(q')]\right\}$$

$$(8.20)$$

where summation over a is performed over atoms in the crystal and $M_a(q) = \langle(q \cdot u_a)^2\rangle/2$. To find the wave field of electrons moving in a vibrating crystal we need to substitute (8.20) in (8.19) and solve (8.19) for $\rho(r, r', E)$.

To evaluate the cross section of backscattering we also need to solve equation (8.19). Note that since the quantum kinetic equation for the density matrix (8.19) is an integral equation (to be compared with the Boltzmann equation (8.3) which is a differential equation in r), instead of defining boundary conditions we need to define the density matrix of the incident electrons $\rho_0(r, r', E)$.

Of course, one should expect that solving equation (8.19) is going to require more effort than solving the Boltzmann equation (8.3). To gain some qualitative understanding of what the solution of (8.19) is going to look like, we start with a model where we need to take into account only *a single* event of phonon scattering through large angles. This model describes the case of *quasielastic* electron backscattering, i.e. the case where we are interested in electrons which lost only a few electronvolts of energy. Since typically electronic excitations give rise to energy losses of the order of 10 eV or more, to a good approximation we may say that quasielastically scattered electrons are those which have undergone *no* interaction with the electronic degrees of freedom of the crystal. Assuming that the mean free path l_{el} of electron–electron interactions is much smaller than the inelastic mean free path l_{ph} of electron–phonon interactions, we may evaluate the cross section of scattering iteratively by substituting ρ_0 in the second term in the right-hand side of equation (8.19). Following a procedure similar to that described by Dudarev and Peng [26] in their study of the angular distribution of quasielastically scattered electrons, we arrive at a simple formula for the number of electrons backscattered in the direction $k_1 = p_1/\hbar$ (here we assume that this direction does not coincide with any of the main crystallographic axes and we may therefore neglect diffraction of backscattered electrons):

$$I(k_0 \to k_1) = N_V l_{\mathrm{el}}\delta\left(\frac{\hbar^2 k_1^2}{2m} - \frac{\hbar^2 k_0^2}{2m}\right) \sum_{m,n} I_{mn}\sigma_{nm}(k_1, k_0) \qquad (8.21)$$

where the cross sections $\sigma_{nm}(k_1, k_0)$ are defined as

$$\sigma_{nm}(k_1, k_0) = \left(\frac{m}{2\pi\hbar^2}\right)^2 \sum_j U_j(k_1 - G_n - k_0) U_j(k_1 - G_m - k_0)$$

$$\times \exp[i(G_n - G_m) \cdot r_j] \left\{ \exp[-M_j(G_n - G_m)] \right.$$

$$\left. - \exp[-M_j(k_0 + G_n - k_1) - M_j(k_0 + G_m - k_1)] \right\}. \tag{8.22}$$

The summation over j in (8.22) goes over atoms in a unit cell and quantities I_{mn} satisfy a closed system of algebraic equations

$$-\frac{1}{l_{el}}\delta_{n0}\delta_{m0} + \frac{I_{nm}}{l_{el}|\cos\theta_1|} + \frac{1}{2l_{el}}\left(\frac{1}{v_n} + \frac{1}{v_m}\right)I_{nm}$$

$$+\frac{i}{\hbar v}\sum_s\left(\frac{H_{ns}}{v_n}I_{sm} - I_{ns}\frac{H_{sm}}{v_m}\right) = 0. \tag{8.23}$$

Here H_{nm} is the Hamiltonian matrix of the diffraction problem

$$H_{nm} = (\epsilon_n - \epsilon_0)\delta_{nm} + U_{nm} \tag{8.24}$$

where U_{nm} are the Fourier components of the (non-Hermitian) crystal potential that include the imaginary part resulting from thermal fluctuations and parameters $\epsilon_n = \hbar^2(k_0 + G_n)^2/2m$ and $v_n = (k_0 + G_n)_z/k_0 > 0$ are responsible for the dependence of the cross section of backscattering on the mutual orientation of the incident beam of electrons and the crystal lattice. In the limiting case where diffraction of both incident *and* backscattered electrons may be neglected, the cross section of scattering acquires a particularly simple form:

$$I(k_0 \rightarrow k_1) = N_V l_{el} \frac{\cos\theta_0 |\cos\theta_1|}{\cos\theta_0 + |\cos\theta_1|}\sigma_{00}(k_1, k_0). \tag{8.25}$$

The latter expression is similar to the one describing electron backscattering from an amorphous material. The effect of the crystal lattice is present in (8.25) in the form of an additional negative term proportional to the Debye–Waller factor, namely

$$\sigma_{00}(k_1, k_0) = \left(\frac{m}{2\pi\hbar^2}\right)^2 \sum_j |U_j(k_1 - k_0)|^2 \left\{1 - \exp[-2M_j(k_0 - k_1)]\right\}.$$

$$\tag{8.26}$$

If we do take diffraction of incident electrons into account, the cross section of backscattering treated as a function of the *direction of incidence* exhibits the behaviour shown in figure 8.4.

Figure 8.4 Electron channelling pattern simulated numerically using the technique described in the text. Darker areas correspond to the directions of incidence where electrons penetrate deeper into the crystal because of the anomalous transmission effect. Brighter areas correspond to the 'anomalous absorption' directions of incidence. This pattern shows the cross section of electron backscattering from a silicon single crystal calculated numerically in the 400-beam approximation for $E_0 = 35$ keV and for the range of angles of incidence $0 \leqslant \phi \leqslant 10°$ and $35° \leqslant \theta \leqslant 55°$.

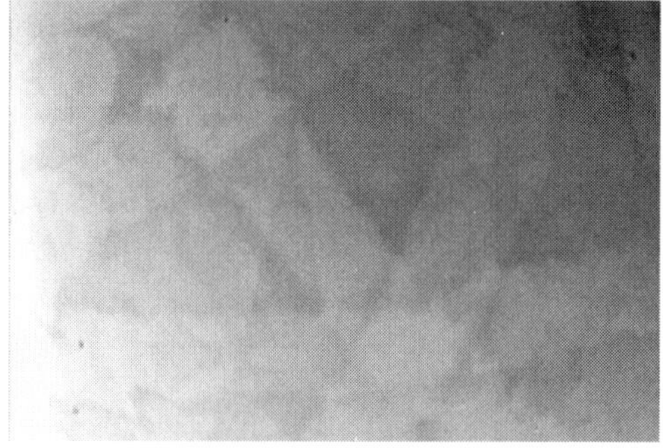

Figure 8.5 Electron channelling pattern showing how the cross section of backscattering depends on the direction of incidence. This pattern was obtained using a copper single crystal at 30 keV for the direction of incidence close to $\langle 111 \rangle$. (Courtesy of J Ahmed.)

The pattern shown in figure 8.4 was calculated numerically using equations (8.21)–(8.24). Figure 8.5 shows how the *experimentally observed* cross section of backscattering depends on the direction of incidence. The dependence of the cross section of electron backscattering on the direction of incidence of the kind shown in figures 8.4 and 8.5 is at present known as an *electron channelling pattern*. This terminology was introduced by Hirsch and Humphreys [27] soon after the experimental discovery by Coates [28] of the fact that the cross section of electron backscattering depends on the orientation of the incident beam of electrons with respect to the crystal lattice. A formula which is similar to (8.21) was obtained by Spencer *et al* [29] who treated electron backscattering using a different approach and studied backward and forward scattering of electrons from a crystal slab.

Clearly, electron channelling patterns represent a manifestation of a very general phenomenon since any cross section of inelastic scattering of high energy electrons by a crystal should be expected to vary as a function of the orientation of the incident beam with respect to the crystal lattice.

However, it turns out that in the problem of electron backscattering several types of scattering contribute in an equally important way to the formation of the observed signal, and this signal itself depends on several factors, for example, on the part of the energy spectrum used for carrying out experimental observations and on the position of the detector with respect to the specimen and the incident beam. To understand all these points in detail we need to develop a method suitable for solving equation (8.19) in a more general case than in the case of quasi-elastic backscattering considered above.

8.5 A NUMERICAL SOLUTION OF THE QUANTUM KINETIC EQUATION

Here we consider a more powerful approach to solving the quantum kinetic equation than the iterative procedure described in the previous section. Indeed, it is known that the iterative approach is only useful if the sum of the first two terms of the relevant series already provides a good approximation to the exact (yet unknown) solution of the problem. In practice this means that iterative methods can only be used if we have a well defined condition determining the rate of convergence of the method. In the case of quasi-elastic backscattering the corresponding inequality has the form $l_{el} \ll l_{ph}$, where l_{el} and l_{ph} are inelastic mean free paths of electron–electron and electron–phonon interactions, respectively. However, the study of various mathematical aspects of the problem of electron backscattering from amorphous materials carried out by Tilinin [10] and by Tilinin and Werner [13] shows that in a more general case the solutions of the Boltzmann equation *do not* contain a (small) parameter which could be employed for representing the cross section of backscattering in a form of a rapidly convergent series.

We are going to follow a different approach to solving the quantum kinetic equation (QKE) for the density matrix (equation (8.19)). We adopt the point of view that since we need to know the distribution of backscattered electrons for a wide range of angles of incidence, it is hard to expect that analytical treatment covering all the aspects of the problem will ever become available. To overcome this difficulty we attempt to develop a numerical approach to solving the QKE. Here I briefly describe the results obtained by Dudarev and Whelan [30] and Dudarev *et al* [31].

We start by stating that to evaluate the dependence of the cross section of electron backscattering on the orientation of the incident beam (or, in other words, to evaluate the contrast of electron channelling patterns) it is necessary to solve an *inhomogeneous* transport equation similar to equation (8.3). The only difference between (8.3) and the equation that needs to be solved to find the probability of electron backscattering from a crystal consists in adding an appropriate source term to the right-hand side of the transport equation. The precise form of the source term in the transport equation can be found by comparing this equation with the QKE (8.19), and a rigorous derivation of the relevant term is given in appendix A of the article by Dudarev *et al* [31]. The idea that it is the *inhomogeneous* transport equation which represents a suitable way of describing electron backscattering from a crystal has been proposed by Spencer and Humphreys [32] who generalized the treatment of multiple scattering of electrons by a crystal developed by Hirsch and Humphreys [27]. The equation which we are now going to study is

$$n\frac{\partial}{\partial r}N(n, E, r) = n_V \int \mathrm{d}o_{n'} \int \mathrm{d}\epsilon \frac{\mathrm{d}^2\sigma}{\mathrm{d}o_{n'}\mathrm{d}\epsilon}(n, E|n', E + \epsilon)N(n', E + \epsilon, r)$$

$$-n_V \int \mathrm{d}o_{n'} \int \mathrm{d}\epsilon \frac{\mathrm{d}^2\sigma}{\mathrm{d}o_{n'}\mathrm{d}\epsilon}(n', E - \epsilon|n, E)N(n, E, r) + Q(n, E, r) \quad (8.27)$$

where $Q(n, E, r)$ describes electrons which have undergone *single* scattering by thermal fluctuations of the crystal potential. It has to be spelled out that equation (8.27) is incomplete. It *does not* describe electrons which have *not* been scattered by phonons and which propagate through the crystal retaining their coherence with the incident beam. To find the density matrix of the *coherently* scattered electrons we have to solve the usual set of equations describing many-beam dynamical diffraction of electrons by the periodic potential of the crystal. The source function in the right-hand side of the transport equation (8.27) has the form

$$Q(n, E, r) = v\left(\frac{m}{2\pi\hbar^2}\right)^2 \delta(E - E_0)N_V \sum_{j,h,l} \phi_h(r)\phi_l^*(r)U_j(k_0 + G_h - k)$$

$$\times U_j(k - k_0 - G_l)\exp[i(G_h - G_l)\cdot r_j]\{\exp[-M_j(G_l - G_h)]$$

$$- \exp[-M_j(k_0 + G_h - k) - M_j(k_0 + G_l - k)]\} \quad (8.28)$$

where summation over j is performed over atoms in a unit cell, E_0 is the energy of the incident electrons, N_V is the number of unit cells per unit volume, v is the velocity of electrons and where functions $\phi_h(x)$ entering (8.28) satisfy a set of Takagi's equations

$$\mathrm{i}\frac{\hbar^2}{m}(k_0 + G_h) \cdot \frac{\partial}{\partial x}\phi_h(x) = (\epsilon_h - \epsilon_0)\phi_h(x) + \sum_t \left(U_{ht} - \tfrac{1}{2}\mathrm{i}\gamma_{ht}\right)\phi_t(x).$$

(8.29)

Together, equations (8.28) and (8.29) satisfy a conservation law

$$\int_V \mathrm{d}^3r \int \mathrm{d}E \int \mathrm{d}o_n\, Q(n, E, r) = -\int_S \mathrm{d}S\left[\frac{\hbar}{m}\sum_h (k_0 + G_h)|\phi_h(r)|^2\right]$$

(8.30)

which shows that the total flux generated by the source function $Q(n, E, r)$ in the transport equation (8.27) equals the decrease of the flux associated with electrons scattered through small angles.

In the case where (i) electron–electron interactions can be treated as effective absorption and (ii) diffraction of incident electrons can be neglected, equation (8.27) acquires a particularly simple form (this of course is an oversimplification which serves the purpose of illustrating the idea of how equation (8.27) can be solved numerically)

$$\cos\theta\,\frac{\partial}{\partial z}N(\cos\theta, \phi, z) = n_V\int_{-\pi}^{\pi}\mathrm{d}\phi'\int_0^{\pi}\mathrm{d}\theta'\,\sin\theta'(\mathrm{d}\sigma[\psi']/\mathrm{d}o')_{\mathrm{el}}N(\cos\theta', \phi', z)$$

$$-n_V\sigma_{\mathrm{tot}}N(\cos\theta, \phi, z) + n_V(\mathrm{d}\sigma[\psi_0]/\mathrm{d}o)_{\mathrm{el}}\exp(-n\sigma_{\mathrm{tot}}z/\cos\theta_0) \qquad (8.31)$$

where $\sigma_{\mathrm{tot}} = \sigma_{\mathrm{el}} + \sigma_{\mathrm{inel}}$. In (8.31) ψ' is the angle between directions of propagation before and after an elastic collision,

$$\cos\psi' = \sin\theta\,\sin\theta'\,\cos(\phi - \phi') + \cos\theta\,\cos\theta' \qquad (8.32)$$

and ψ_0 is given by (8.32) with θ' replaced by θ_0 and ϕ' replaced by ϕ_0, θ_0 and ϕ_0 being the angles of incidence (see figure 8.2 which illustrates the geometry of scattering). The boundary condition on (8.31) at $z = 0$ has the form

$$N(\cos\theta, \phi, z = 0) = 0 \quad \text{for} \quad \cos\theta > 0 \qquad (8.33)$$

and our goal is to determine the coefficient of backscattering

$$R_{\mathrm{tot}} = \int_{-\pi}^{\pi}\mathrm{d}\phi\int_{\pi/2}^{\pi}\mathrm{d}\theta\,\sin\theta|\cos\theta|N(\cos\theta, \phi, z = 0). \qquad (8.34)$$

To solve equation (8.31) numerically we introduce a dimensionless variable $\tau = n_V \sigma_{tot} z$ and approximate the integral term by an L-point (where L equals 48 or 96) Gaussian quadrature. We obtain

$$\mu_j \frac{\partial}{\partial \tau} N(\mu_j, \tau) = \frac{1}{w_{tot}} \sum_{j'=1}^{L} \Phi(\mu_j, \mu_{j'}) N(\mu_{j'}, \tau) W_{j'} - N(\mu_j, \tau)$$

$$+ \frac{1}{w_{tot}} \Phi(\mu_j, \mu_0) \exp(-\tau/\mu_0) \tag{8.35}$$

where $w_{tot} = n_V \sigma_{tot}$, $\mu = \cos\theta$, $\mu_0 = \cos\theta_0$ and

$$\Phi(\mu, \mu') = n_V \int_{-\pi}^{\pi} d\phi \, d\sigma_{el} \big[\cos^{-1}\{(1 - \mu^2)^{1/2}(1 - \mu'^2)^{1/2} \cos\phi + \mu\mu'\} \big] / do'.$$

$$\tag{8.36}$$

The system of linear differential equations (8.35) represents an approximate form of the transport equation (it coincides with the transport equation in the limit $L \to \infty$), and it can be easily solved by finding the eigenvalues and eigenvectors of the matrix $\Pi_{jj'} = (w_{tot}\mu_j)^{-1}[\Phi(\mu_j, \mu_{j'})W_{j'} - \delta_{jj'}]$. Since the flux of electrons considered as a function of coordinate z in the direction normal to the surface must vanish in the limit $z \to \infty$, the solution of (8.35) must contain no terms diverging at $z = \infty$. This condition leads to an explicit expression for the coefficient of backscattering

$$R_{tot} = \sum_{j=1}^{L} \sum_{m=1}^{L} |\mu_j| W_j X_{jm} C_m(0) \tag{8.37}$$

where W_j denote Gaussian weights, X_{jm} are the eigenvectors of $\Pi_{jj'}$ and $C_m(0)$ are the coefficients involved in the expansion

$$N(\mu_j, \tau) = \sum_{m=1}^{L} X_{jm} C_m(\tau) \exp(p_m \tau). \tag{8.38}$$

The numerical implementation of the procedure described above is straightforward. The comparison of total coefficients of quasielastic backscattering calculated using equations (8.31)–(8.38) with experimental data obtained by Schmid *et al* [33] shows that theoretical and experimental values agree within 10%. For example, for polycrystalline copper at $E_0 = 1$ keV we find $R = 3.3\%$ (this value was evaluated for the geometry of scattering described by Schmid *et al* [33]) while the experimentally observed probability of quasi-elastic backscattering equals 3.4%.

8.6 THE ROLE OF MULTIPLE SCATTERING AND ENERGY LOSSES

In the previous section we showed how an inhomogeneous transport equation can be solved numerically in the case of quasi-elastic electron backscattering. This solution demonstrates that straightforward numerical integration of equation (8.27) is indeed possible, but for practical purposes a more general approach needs to be developed. The main point of this development concerns the treatment of energy losses.

It may sound somewhat strange, but the fact is that *energy losses* are entirely responsible for the variation of the *total* coefficient of electron backscattering from a crystal. In other words, the contrast of electron channelling patterns observed using the mode where we look at the variation of the total coefficient of backscattering R_{tot} treated as a function of the two angles of incidence ϕ_0 and θ_0 depends entirely on how electrons lose energy as they scatter inside the crystal and escape back through its surface into the vacuum.

It is a simple matter to understand the validity of the above statement. Indeed, let us assume that in a solid electrons are scattered in all directions by thermal fluctuations of the potential but that they do not lose energy. In this case, given a sufficiently long interval of time, any electron incident on the surface will eventually approach it from inside and cross it, leaving the crystal forever. In other words, if the beam of incident electrons remains stationary, at any given moment of time the number of particles entering the crystal equals the number of particles leaving it. This statement evidently does not depend on the choice of the direction of incidence and shows that in the absence of energy losses the total coefficient of backscattering should remain constant, $R_{\text{tot}} = 1$, independently of the mutual orientation of the incident beam and the crystal lattice.

A full-scale treatment of the problem of electron backscattering therefore requires solving an equation which takes into account both incoherent quasielastic scattering of electrons by thermal fluctuations of the potential *and* electron energy losses. This equation has the form

$$\cos\theta \frac{\partial}{\partial z} N(\cos\theta, \phi, E, z) = \int_{-\pi}^{\pi} d\phi' \int_{0}^{\pi} d\theta' \sin\theta' w_{\text{ph}}(\cos\psi', E) N(\cos\theta', \phi', E, z)$$

$$+ \frac{\partial}{\partial E}[\bar{\epsilon}(E) N(\cos\theta, \phi, E, z)]$$

$$- w_{\text{ph}}^{(\text{tot})}(E) N(\cos\theta, \phi, E, z) + Q(\cos\theta, \phi, E, z) \tag{8.39}$$

where the subscript 'ph' refers to the cross section of phonon scattering as opposed to the cross section of energy losses represented by the second term in the right-hand side of equation (8.39). The cross section of energy losses in this equation is written using the continuous slowing down approximation. In this approximation the rate of energy losses is assumed to be independent of the direction of motion of the electron and the amount of energy lost by a given electron is assumed to be proportional to the length of the trajectory along which this electron travelled from the point where it had entered the crystal.

The boundary condition for equation (8.39) has the form

$$N(\mu, E, z = 0) = 0 \quad \text{for} \quad 0 < \mu < 1 \tag{8.40}$$

and we are interested in finding the distribution of electrons $N(\mu, E, z = 0)$ in the region of negative values of μ, namely, in the region $-1 < \mu < 0$.

Is it possible to apply the method developed for solving equation (8.31) directly to equation (8.39)? A simple estimate shows that this is going to be very demanding computationally and will require the diagonalization of a very large matrix of the order of 3000×3000. In order to avoid passing through this time consuming computational step, in solving equation (8.39) we shall rely on the supermatrix algorithm proposed by Fathers and Rez [34, 35]. The main idea of this method consists in using the fact that according to equation (8.39) the term describing energy losses can be approximated by a finite difference

$$\frac{\partial}{\partial E}[\bar{\epsilon}(E)N(\cos\theta, \phi, E, z)] \approx \frac{\bar{\epsilon}(E_{i-1})}{\Delta E}N(\cos\theta, \phi, E_{i-1}, z)$$

$$-\frac{\bar{\epsilon}(E_i)}{\Delta E}N(\cos\theta, \phi, E_i, z) \tag{8.41}$$

so that the entire energy spectrum may then be interpolated through a finite set of points $E_0 > E_1 > E_2, \ldots$. In this case the process of the electron losing energy can be represented by a series of successive transitions of electrons from higher to lower energy levels in accordance with equation (8.41). In practice, solving equation (8.39) involves several additional transformations, for example, expanding $N(\cos\theta, \phi, E, z)$ in the Fourier series in ϕ. The main computational step consists in the successive diagonalization of matrices describing phonon scattering of electrons *at a given energy* followed by the recursive evaluation of angular distribution of backscattered electrons at every point $E_0, E_1, E_2, \ldots$ of the energy spectrum.

To illustrate numerical solutions of equation (8.39), found using the supermatrix algorithm, we consider how the shape of the energy spectrum of backscattered electrons depends on the direction of the incident beam, or, in other words, how this spectrum varies as a function of the mutual orientation of the incident beam and the crystal lattice.

Figure 8.6 shows the energy spectrum of electrons backscattered from a silicon single crystal. The maximum of the spectrum is shifted slightly down from the energy E_0 of the incident electrons, and this represents an interesting feature of the solution obtained using the supermatrix algorithm (this maximum can indeed be readily observed experimentally, see Wolf *et al* [36], and it is

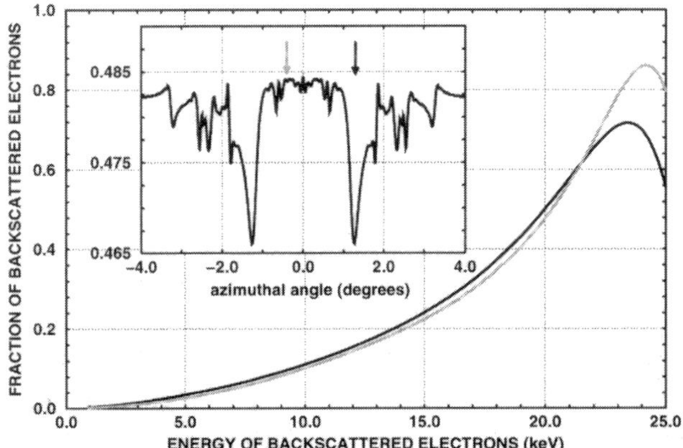

Figure 8.6 Energy spectra of electrons backscattered from a silicon single crystal. The spectrum shown in grey corresponds to the direction of incidence where the total coefficient of backscattering is a maximum (the direction of incidence corresponds to the enhanced anomalous absorption of electrons). The spectrum shown in black corresponds to the direction of incidence where the total coefficient of backscattering is a minimum (the direction of anomalous transmission of electrons). The total area under the curves is proportional to the total coefficient of backscattering corresponding to a given direction of incidence. Energy E_0 of incident electrons equals 25 keV and the angle of incidence θ_0 is equal to 69°. The dependence of the total coefficient of backscattering on the azimuthal angle ϕ_0 is shown in the inset.

absent in solutions obtained using less accurate approximations, see for example Sandström *et al* [37]).

Another experimentally observed effect, the origin of which can now be understood on the basis of the numerical solution of equation (8.39), is the phenomenon of the contrast reversal of channelling patterns. This effect was discovered by Ichinokawa *et al* [38] who showed that the *sign* of the contrast of electron channelling patterns depends on the position of the detector of backscattered electrons. The nature of this phenomenon is associated with the change of the shape of the *angular distribution* of electrons backscattered from a crystal at grazing incidence with the change of the direction of incidence. To explain the origin of this effect theoretically, it is necessary to take into account the interplay between anomalous transmission effects, energy losses and *multiple* scattering of electrons by thermal fluctuations of the potential. This can be achieved by solving equation (8.39) numerically as shown by Dudarev, Rez and Whelan [31].

8.7 A PERTURBATION TREATMENT OF ECC IMAGES

So far we have been discussing electron backscattering by a perfect crystal. We showed that to develop a consistent theoretical approach to electron backscattering one needs to be able to treat dynamical diffraction of electrons, multiple scattering of electrons by phonons and electron energy losses on an equal footing, and this can be achieved using the quantum kinetic equation for the density matrix. However, for practical purposes it is insufficient to be able to treat backscattering by a perfect crystal. Experimentally it is now possible to use backscattered electrons to obtain images of defects (in particular, images of dislocations) emerging on the surface from the crystal bulk or situated sufficiently close to the surface of the crystal.

For the first time the idea that channelling of electrons could be used to obtain images of defects in backscattering mode was proposed by Booker *et al* [39]. First calculations of lattice defect images were performed by Clarke and Howie [40]. Later Clarke [41] made an attempt to observe crystal defects in SEM using backscattered electrons. To improve the contrast of SEM images of defects, Morin *et al* [42] performed a series of experiments in which they showed how electron channelling images of various types of defect can be obtained employing energy filtering and collecting only the quasielastically scattered electrons. More recently, Czernuszka *et al* [43], Wilkinson *et al* [44, 45, 5] and Ahmed *et al* [46] demonstrated that it is actually possible to obtain high quality electron channelling contrast images (ECCIs) of defects using a commercial SEM and employing no energy filtering. This has been achieved using the inclined geometry of scattering and a specially designed highly sensitive detector of backscattered electrons. An example of an electron channelling contrast image of fatigue dislocation structures (persistent slip band) in a single crystal of copper is shown in figure 8.7.

To simulate an energy-unfiltered electron channelling contrast image of a defect we follow the idea proposed by Howie [47] who suggested that since the observed level of contrast is relatively low (normally it is of the order of 10%) it should be possible to develop a *perturbative* approach to the calculation of the *variation* of the coefficient of backscattering associated with bending of atomic planes around the defect. The formal treatment of the problem follows the standard 'quantum-mechanical' procedure of transforming the transport equation into the integral form, introducing the Green's function of the transport equation and solving it iteratively. The first-order term of this series has the form

$$\delta R(\cos\theta, \phi, E) = R(\cos\theta, \phi, E) - R_{\text{rand}}(\cos\theta, \phi, E)$$

$$\sim \int \mathrm{d}S \int_0^\infty \mathrm{d}z\, z \left[\sum_{h,l} \Phi_h(r)\Phi_l^*(r)\frac{\gamma_{lh}}{\hbar} - v_0 w_{\text{el}}^{(\text{tot})}(E_0) \exp\left(-\frac{w_{\text{el}}^{(\text{tot})}(E_0)z}{\cos\theta_0} \right) \right]$$

$$(8.42)$$

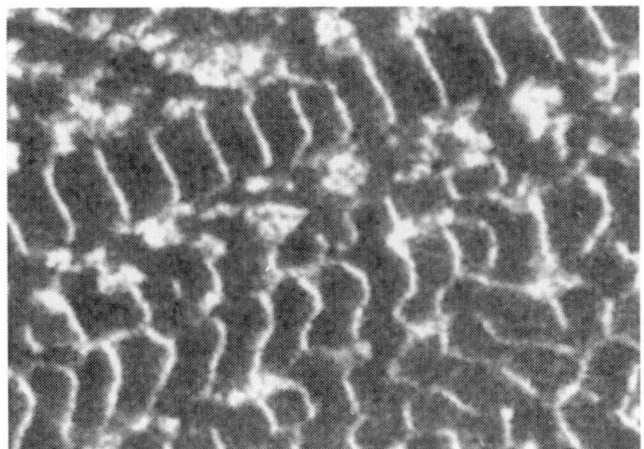

Figure 8.7 Electron channelling contrast image of a fatigued copper single crystal at saturation. A typical ladder structure of a persistent slip band which consist of dislocation walls is seen in this image. This image was obtained at the Department of Materials, University of Oxford using a JEOL 6300 SEM operated at 30 keV at the level of magnification of 3000 times. The image shown in this figure was obtained using the outer dark edge of the (220) channelling band. (Courtesy of J Ahmed.)

where v_0 is the velocity of the incident electrons, $w_{\mathrm{el}}(\cos\psi, E)$ is proportional to the differential cross section of phonon scattering and θ_0 and ϕ_0 are the polar and azimuthal angles of incidence. The coefficients γ_{lh} are related to the imaginary part of the Fourier components of the crystal potential via $\gamma_{hl} = \mathrm{i}[U_{hl} - (U)_{hl}]$. Amplitudes $\Phi_h(\boldsymbol{r})$ entering (8.42) satisfy the system of coupled linear differential equations

$$\mathrm{i}\frac{\hbar^2}{m}(\boldsymbol{k}_0 + \boldsymbol{G}_h)\frac{\partial}{\partial \boldsymbol{r}}\Phi_h(\boldsymbol{r}) = -\Phi_h(\boldsymbol{r})\frac{\hbar^2}{m}\left\{(\boldsymbol{k}_0 + \boldsymbol{G}_h)\cdot\frac{\partial}{\partial \boldsymbol{r}}\right\}(\boldsymbol{G}_h\cdot\boldsymbol{R}(\boldsymbol{r}))$$

$$+[\epsilon_h - \epsilon_0]\Phi_h(\boldsymbol{r}) + \sum_t U_{ht}\Phi_t(\boldsymbol{r}) \tag{8.43}$$

where the Fourier components of the potential U_{ht} coincide with those of the undistorted lattice and the boundary condition on (8.43) at $z = 0$ is

$$\Phi_h(\boldsymbol{r})|_{z=0} = \delta(x - x_0)\delta(y - y_0)\delta_{h0}$$

where (x_0, y_0) are the coordinates of the point on the surface where the incident beam enters the crystal.

Results of theoretical simulation of ECC images of an inclined dislocation based on direct numerical integration of (8.43) for a general dislocation characterized by the displacement field $\boldsymbol{R}(\boldsymbol{r})$ given by Shaibani and Hazzledine [48] are shown in figure 8.8. Contrast variations seen in the simulated images have indeed been observed in both energy-filtered (Morin *et al* [42]) and

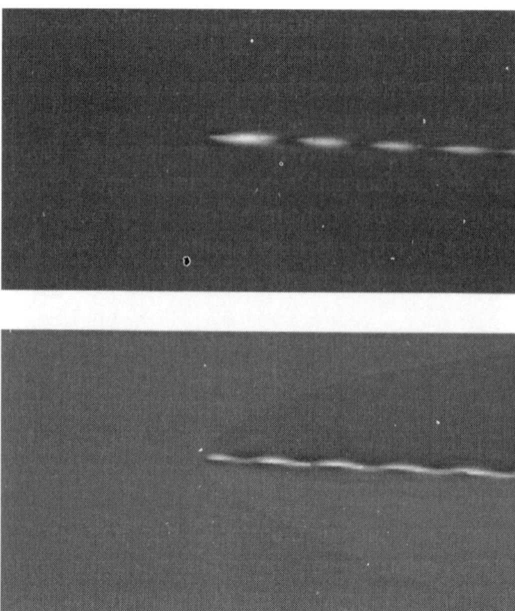

Figure 8.8 Simulated channelling contrast images of 60° inclined misfit dislocations in Si. The Burgers vector b of the dislocation makes the angle of 60° or 120° with the [01$\bar{1}$] direction. Top: ECCI simulated for $b\|[110]$. Bottom: ECCI simulated for $b\|[\bar{1}10]$. See Dudarev *et al* [49] for more detail.

in energy-unfiltered ECCIs (J T Czernuszka and A J Wilkinson, private communication) at sufficiently high magnification.

Experimental images of inclined dislocations similar to those shown in figure 8.8 have been obtained by Czernuszka *et al* [43]. It is likely that a number of features characterizing defects situated near the surface can be analysed by using equations (8.42) and (8.43) and by comparing simulated images with experimental observations. At the same time the question of how to simulate and to interpret images of arrays of dislocations and dislocation walls of the kind shown in figure 8.7 still represents a challenge for the theory (very recently this point has been addressed by Dudarev, Ahmed, Hirsch and Wilkinson [50]). To explain the origin of the contrast features seen in figure 8.7 it is necessary both to understand the geometry of dislocation lines forming fatigue structures in the crystal and to describe various types of scattering process contributing to the formation of the image. The fascinating story of diffraction effects in electron backscattering continues. . ..

8.8　SUMMARY

This chapter represents an attempt to give a short survey of the current state of theoretical understanding of the problem of electron backscattering from a crystalline material and its relation to the state-of-the-art experimental work. It is shown how, by starting from the transport equation describing scattering of high energy electrons by an amorphous material and by introducing the quantum kinetic equation for the density matrix of high energy electrons, it is possible to arrive at a comprehensive and numerically manageable mathematical treatment of electron backscattering from a crystalline solid. Analytical and numerical results discussed above illustrate how the development of a consistent approach to the problem makes it possible to explain the origin of a number of interesting and somewhat mysterious effects observed experimentally. It is also shown how methods developed for calculating the contrast of electron channelling patterns can be used to simulate electron channelling contrast images of crystal defects.

ACKNOWLEDGMENTS

I would like to acknowledge stimulating discussions with Sir Peter Hirsch, M J Whelan, C J Humphreys and P Rez. I thank Fellows of Linacre College, Oxford, for the provision of a Research Fellowship. Computations were performed in the Materials Modelling Laboratory of the Department of Materials at the University of Oxford.

REFERENCES

[1]　Niedrig H 1982 *J. Appl. Phys.* **53** R15

[2]　Joy D C, Newbury D E and Davidson D L 1982 *J. Appl. Phys.* **53** R81

[3]　Booker G R 1970 *Modern Diffraction and Imaging Techniques in Materials Science* ed S Amelinckx, R Gevers, G Remault and J Van Landuyt (Amsterdam: North-Holland) pp 557, 597, 613

[4]　Dudarev S L 1997 *Micron* **28** 139

[5]　Wilkinson A J and Hirsch P B 1997 *Micron* **28** 279

[6]　Kalashnikov N P, Remizovich V S and Ryazanov M I 1980 *Collisions of Fast Charged Particles in Solids* (Moscow: Atomizdat) p 7

[7]　Ryazanov M I and Tilinin I S 1985 *Surface Analysis Using Backscattered Particles* (Moscow, Energoatomizdat) p 35

[8]　Abrikosov A A, Gorkov L P and Dzyaloshinski I E 1975 *Methods of Quantum Field Theory in Statistical Physics* (New York: Dover) p 330

[9]　Wang M C and Guth E 1951 *Phys. Rev.* **84** 1092

[10]　Tilinin I S 1982 *Sov. Phys. –JETP* **55** 751

[11]　Tilinin I S and Werner W S M 1992 *Phys. Rev.* B **46** 13 739

[12] Tilinin I S and Werner W S M 1993 *Surf. Sci.* **290** 119
[13] Tilinin I S and Werner W S M 1994 *Mikrochim. Acta* **114/115** 485
[14] Landau L D and Lifshits E M 1977 *Quantum Mechanics, Non-Relativistic Theory* 3rd edn (Oxford: Pergamon)
[15] Chandrasekhar S 1950 *Radiative Transfer* (Oxford: Clarendon)
[16] Boersch H, Wolter R and Shoenebeck H 1967 *Z. Physik* **199** 124
[17] Boersch H 1971 *Proc. 25th Ann. Meeting EMAG* (*Inst. Phys. Conf. Ser. 10*) ed W C Nixon (London: Institute of Physics) p 50
[18] Makarov V V, Artem'ev V P, Igonin S I and Petrov N N 1986 *Problems of Physical Electronics* (Leningrad: A F Ioffe Physical and Technical Institute) p 741
[19] Kagan Yu M 1962 *Mössbauer Effect* (Moscow, IIL) p 3
[20] Igonin S I and Makarov V V 1987 *Sov. Phys.–Tech. Phys. Lett.* **13** 1043
[21] Frisch U 1968 *Probabilistic Methods in Applied Mathematics* ed A T Bharucha-Reid (New York: Academic) p 75
[22] Mahan G D 1990 *Many-Particle Physics* 2nd edn (New York: Plenum) p 855
[23] Gorodnichev E E, Dudarev S L, Rogozkin D B and Ryazanov M I 1987 *Sov. Phys.–JETP* **66** 938
[24] Lifshits E M and Pitaevskii L P 1979 *Physical Kinetics* (Oxford: Pergamon) 95
[25] Dudarev S L, Peng L M and Whelan M J 1993 *Phys. Rev.* B **48** 13 408
[26] Dudarev S L and Peng L M 1991 *Surf. Sci.* **244** L133
[27] Hirsch P B and Humphreys C J 1970 *Scanning Electron Microscopy 1970* ed O Johari (Chicago, IL: IIT Research Institute) p 451
[28] Coates D G 1967 *Phil. Mag.* **16** 1179
[29] Spencer J P, Humphreys C J and Hirsch P B 1974 *Phil. Mag.* **26** 193
[30] Dudarev S L and Whelan M J 1994 *Surf. Sci.* **311** L687
[31] Dudarev S L, Rez P and Whelan M J 1995 *Phys. Rev.* B **51** 3397
[32] Spencer J P and Humphreys C J 1980 *Phil. Mag.* **42** 433
[33] Schmid R, Gaukler K H and Seiler H 1983 *Scanning Electron Microscopy*, ed O Johari (Chicago, IL: SEM) pt II, p 501
[34] Fathers D J and Rez P 1979 *Scanning Electron Microscopy 1979* ed O Johari (Chicago, IL: SEM) pt I, p 55
[35] Fathers D J and Rez P 1984 *Electron Beam Interactions with Solids*, ed D F Kyzer, H Niedrig, D E Newbury and R Shimizu (Chicago, IL: SEM) p 193
[36] Wolf E D, Coane P J and Everhart T E 1970 *Microscopie Electronique 1970* vol II, ed P Favard (Paris: French Society of Electron Microscopy) p 595
[37] Sandström R, Spencer J F and Humphreys C J 1974 *J. Phys. D: Appl. Phys.* **7** 1030
[38] Ichinokawa T, Nishimura M and Wada H 1974 *J. Phys. Soc. Japan* **36** 221

[39] Booker G R, Shaw A M B, Whelan M J and Hirsch P B 1967 *Philos. Mag.* **16** 1185

[40] Clarke D R and Howie A 1971 *Phil. Mag.* **24** 959

[41] Clarke D R 1971 *Phil. Mag.* **24** 973

[42] Morin P, Pitaval M, Besnard D and Fontaine G 1979 *Phil. Mag.* **40** 511

[43] Czernuszka J T, Long N J, Boyes E D and Hirsch P B 1990 *Phil. Mag. Lett.* **62** 227

[44] Wilkinson A J, Anstis G R, Czernuszka J T, Long N J and Hirsch P B 1993 *Phil. Mag.* **68** 59

[45] Wilkinson A J 1996 *Phil. Mag. Lett.* **73** 337

[46] Ahmed J, Wilkinson A J and Roberts S G 1997 *Phil. Mag. Lett.* **76** 237

[47] Howie A 1974 *Quantitative Scanning Electron Microscopy* ed D B Holt, M D Muir, P R Grant and I M Boswarva (London: Academic) p 183

[48] Shaibani S J and Hazzledine P M 1981 *Phil. Mag.* **44** 657

[49] Dudarev S L, Czernuszka J T, Peng L M, Wilkinson A J and Whelan M J 1995 *Microscopy of Semiconducting Materials 1995 (Inst. Phys. Conf. Ser. 146)* ed A G Cullis and A E Staton-Bevan (Bristol: Institute of Physics) p 763

[50] Dudarev S L, Ahmed J, Hirsch P B and Wilkinson A J 1999 *Acta Crystallogr.* A **55** 234

9

DEVELOPMENTS OF DYNAMICAL THEORY OF RHEED AND APPLICATIONS TO THE *IN SITU* MONITORING OF MBE GROWTH

Lian-Mao Peng

9.1   INTRODUCTION

On their very discovery of the phenomenon of electron diffraction, Davisson and Germer [1] employed a diffraction geometry which is basically a reflection electron diffraction geometry and considered the reflected beam intensities as a function of the primary beam energy. Nowadays the technique evolved from their early experimental set-up is usually called *low energy electron diffraction* (LEED) [2] to distinguish it from the main subject of this chapter—*reflection high energy electron diffraction* (RHEED).

In RHEED experiments high energy electrons (10 keV and above) are incident on the surface of a crystal with a glancing angle which ranges from one to several degrees (see figure 9.1(*a*)). By selecting one of the reflected beams from the surface, usually the specular reflected beam, a reflection electron microscopy (REM) image showing such atomic details as monatomic surface steps and a single dislocation may be formed (see figure 9.1(*b*)). Although the great majority of known surface structures are determined by LEED [3], there are many reasons why RHEED is still an attractive technique for surface studies. On the one hand the glancing angle incidence makes the technique extremely surface sensitive and an ideal technique for combination with molecular-beam epitaxial (MBE) growth techniques for surface and growth studies [4]. On the other hand the use of high energy electrons simplifies the theoretical treatment of diffraction processes. This is because for high energy electron diffraction the complicated exchange and correlation effects (see, for example, Pendry [2]) are negligible

(*a*)
(*b*)

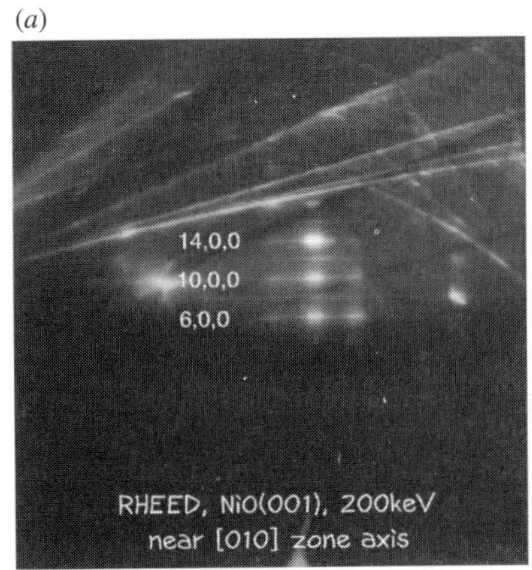

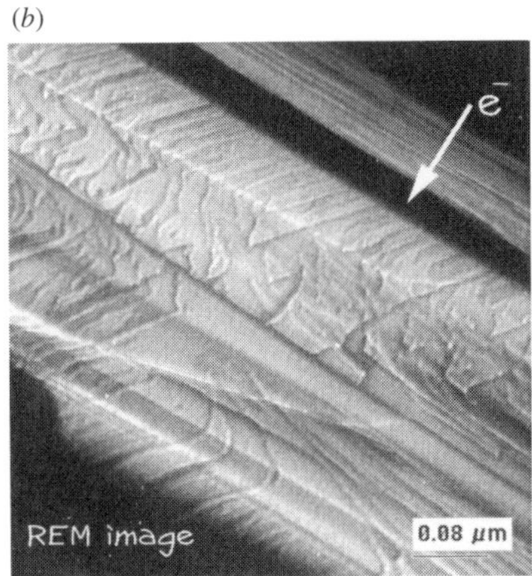

Figure 9.1   Experimental (*a*) RHEED pattern and (*b*) corresponding
REM image taken from the (001) surface of a single crystal of NiO
for a primary beam energy of 200 keV and a beam azimuth near the
[010] zone axis.

and the movement of high energy electrons in a solid is basically governed by
the electrostatic potential between the high energy electrons and nuclei of the
solid. The scattering potential may therefore be constructed readily (see, for
example, Cowley [5]) and calculations of RHEED intensities should be both
easier and faster than equivalent calculations for LEED.

The first dynamical theory of electron diffraction was developed by Bethe [6]
in order to account quantitatively for the experimental observations of Davisson
and Germer [1]. In particular his introduction of the concept of inner potential
had proved to be of crucial importance to the correct indexing of the Davisson
and Germer experiments. In his original theory, Bethe assumed a triply periodic
potential and treated the potential as though it ceases suddenly at a mathematical
plane. Shortly after the publication of Bethe's 1928 paper, von Laue [7] pointed
out that this simplified treatment is in complete contradiction to electrostatics.
By treating the atoms as point charges, von Laue estimated the influence of the
gradual transition between the potential field exterior to and in the interior of
a crystal on the diffraction of electrons. He concluded that this influence may
be neglected for electrons whose energy is 200 eV or more, and one can then
treat the transition as discontinuous. In the first two sections of this chapter we
will first present a general matrix representation of the Bethe theory and its
extension to include gradual transition of the potential field from the interior of
a crystal through a transition surface region, which may or may not have the
same structure as in the interior of the crystal, to the vacuum region exterior to
the crystal. We will show that even for electrons of energy as high as 200 keV
the influence due to the gradual transition is not negligible. A general procedure

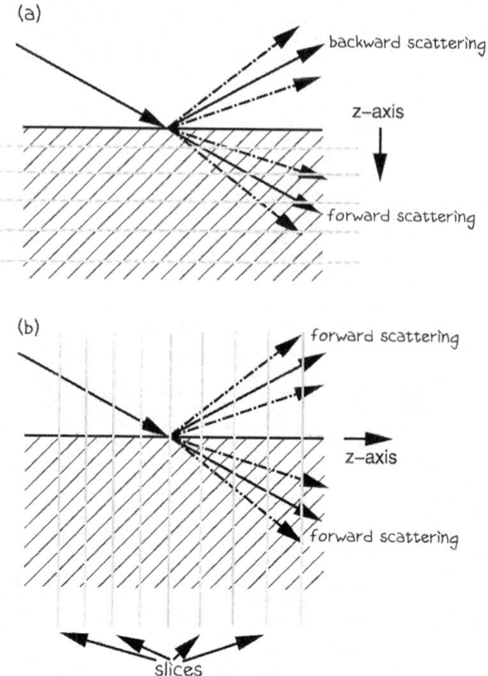

Figure 9.2 Schematic diagram showing the two coordinate systems for dynamical RHEED calculations. In (a) both forward and backscattering from the surface of a crystal are present and in (b) the diffraction processes are basically forward scattering.

for performing many-beam dynamical RHEED calculations will be discussed in detail and illustrated via worked examples.

It should be pointed out that in all theoretical treatments of RHEED, except the one by Peng and Cowley [8, 9], the crystal surface is taken to be the plane separating forward and backward scattering (see figure 9.2(a)). Since in RHEED the electrons are incident on the surface with a glancing angle, both forward and backward scattering are important and must be taken into account. The major difficulty for theoretical treatment of RHEED results from the fact that the inclusion of backscattering produces a set of non-linear differential equations or equivalently a second order eigensystem. Most methods of dynamical RHEED calculations based on this coordinate system differ mainly by their ways of linearization of the basic equations of the dynamical theory of electron diffraction, or the Bethe equations. Alternatively the plane separating forward and backward scattering may be taken to be perpendicular to the surface of the crystal (see figure 9.2(b)). Diffraction processes are then dominated by forward diffraction processes and in principle all numerical procedures for transmission electron diffraction (TED) calculations may be employed for RHEED calculations. In the present chapter we will be mainly concerned with the Bloch wave approach to RHEED. Interested readers are referred to the original papers of Peng and Cowley [8, 9] and some more recent developments [10–12].

A detailed knowledge of the surface structure evolution of a crystal is of the utmost importance in understanding both the physical and chemical properties of surfaces and growth of crystals. Among the various techniques which are capable of providing structural information down to atomic level (see for example [13]), RHEED has the unique advantage of being the most powerful and versatile technique for growth and surface studies of thin films prepared by molecular beam epitaxy (MBE) [14]. In particular, the technique of RHEED intensity oscillations has been used routinely to calibrate beam fluxes and alloy compositions and to control the thicknesses of quantum wells and superlattice layers (see the many articles in [4]). In the last section of this chapter we will consider methods for describing RHEED from rough surfaces, in particular MBE growing surfaces, and to discuss the origin of the phenomenon of RHEED intensity oscillations and its applications to the *in situ* monitoring and controlling of MBE growth of thin films.

9.2 BETHE THEORY—THREE-DIMENSIONAL BLOCH WAVES

In Bethe's original theory [6], the crystal potential is assumed to be three dimensional periodic in space and the electrostatic field ceases suddenly at a certain mathematical plane of zero thickness. While this approach may serve as a good starting point for the description of the diffraction processes of high energy electrons within the bulk crystal, the treatment of the dynamical diffraction processes occurring at the near surface region is not equally satisfactory. In this section we will first discuss Bethe's original theory and in the following section consider how the theory may be generalized to include the treatment of the smooth transition of the electrostatic field from the interior of a crystal into the zero field in the vacuum.

The Bethe theory starts by plane wave expansion of both the electron wave function

$$b(\boldsymbol{k}, \boldsymbol{r}) = \sum_{g} C_g \exp\{i\boldsymbol{k}_g \cdot \boldsymbol{r}\}$$

and the scattering potential $U(\boldsymbol{r}) = 2\,meV(\boldsymbol{r})/\hbar^2$, $V(\boldsymbol{r})$ being the conventional electrostatic potential of the crystal,

$$U(\boldsymbol{r}) = \sum_{g} U_g \exp\{i\boldsymbol{g} \cdot \boldsymbol{r}\}$$

in which g is one of the reciprocal lattice vectors, and $k_g = k+g$ is the diffracted wave vector associated with g. Substituting the above expansions into the wave equation we then obtain the famous Bethe equations (see for example [15, 16]) or the basic equations of dynamical electron diffraction

$$[K^2 - (k+g)^2]C_g + \sum_{h \neq g} U_{g-h}C_h = 0 \tag{9.1}$$

where K is the electron wave vector derived from k_0, the incident plane wave vector, after correction for the mean inner potential, and K^2 is given by $K^2 = k_0^2 + U_0$.

The set of equations (9.1) are indeed second-order eigenvalue equations. Since in the processes of dynamical electron diffraction by crystals the surface parallel component of the electron wave vector k is a good quantum number, only the surface normal component of the wave vector k_z is left with some freedom. The general three-dimensional eigensystem (9.1) can therefore be reduced to a one-dimensional eigensystem by introducing an eigenvalue γ

$$k = k_0 + \gamma n \tag{9.2}$$

where n is a unit vector pointing toward the positive z direction. The set of equations (9.1) then becomes

$$\{\gamma^2 + 2(k_0 + g)_z \gamma - 2k_0 S_g\}C_g - \sum_{h \neq g} U_{g-h}C_h = 0 \tag{9.3}$$

where

$$2k_0 S_g = K^2 - (k_0 + g)^2 \tag{9.4}$$

and S_g is the usual excitation error for reflection g measuring the distance between the reflection g and the Ewald sphere. In matrix notation, equation (9.3) can be re-written as

$$(\gamma^2 \mathbf{I} + \gamma \mathbf{D} + \mathbf{Q})C = 0 \tag{9.5}$$

in which $\mathbf{I}$ is an identity matrix, $\mathbf{D}$ is a diagonal matrix with

$$\{\mathbf{D}\}_{gg} = 2(k_0 + g)_z \tag{9.6}$$

and $\mathbf{Q}$ is a general matrix whose elements are

$$\{\mathbf{Q}\}_{gh} = -2k_0 S_g \delta_{gh} - U_{g-h}(1 - \delta_{gh}) \tag{9.7}$$

where δ_{gh} is the Kronecker δ, and C is a column vector with

$$\{C\}_h = C_h. \tag{9.8}$$

Mathematically, the matrix equation (9.5) is only a special case of a general high degree eigenvalue problem

$$(\gamma^m C_0 + \gamma^{m-1} C_1 + \cdots + \gamma C_{m-1} + C_m) X = 0 \tag{9.9}$$

which can be solved by forming two super-matrices **A** and **B**:

$$\mathbf{A} = \begin{pmatrix} -C_1 & -C_2 & -C_3 & \cdots \\ \mathbf{I} & 0 & 0 & \cdots \\ 0 & \mathbf{I} & 0 & \cdots \\ \cdots & \cdots & \cdots & \cdots \end{pmatrix} \quad \text{and} \quad \mathbf{B} = \begin{pmatrix} C_0 & 0 & 0 & 0 \\ 0 & \mathbf{I} & 0 & 0 \\ 0 & 0 & \mathbf{I} & 0 \\ \cdots & \cdots & \cdots & \cdots \end{pmatrix}$$

such that the original high degree problem (9.9) is transformed into a first order problem

$$\mathbf{A} Z = \gamma \mathbf{B} Z \tag{9.10}$$

where

$$Z^T = (\gamma^{m-1} X, \gamma^{m-2} X, \ldots, X)$$

in which the superscript T denotes transposition of the vector. The first order linear system (9.10) can then be solved using standard numerical routines, such as NAG [17] and EISPACK [18] routines.

For dynamical electron diffraction, $m = 2$, equation (9.5) becomes

$$\begin{pmatrix} -\mathbf{D} & -\mathbf{Q} \\ \mathbf{I} & 0 \end{pmatrix} \begin{pmatrix} \gamma C \\ C \end{pmatrix} = \gamma \begin{pmatrix} \gamma C \\ C \end{pmatrix} \tag{9.11}$$

and this is the eigenvalue equation first introduced by Colella [19] for solving RHEED problems. For cases when there exist a limited number of N reciprocal lattice points lying on n rods of a reciprocal lattice (with rod here we mean those points with the same surface parallel but different z components) and close to the Ewald sphere, only N diffracted beams will be appreciably excited in the crystal. The set of infinite equations (9.11) may then be truncated into a set of $2N$ equations, giving in general $2N$ values of $\gamma^{(j)}(j = 1, \ldots, 2N)$ and $2N$ associated Bloch waves $b^{(j)}(k^{(j)}, r)(j = 1, \ldots, 2N)$.

Having solved the eigenvalue equation (9.11) for N beams, the total electron wave function within the crystal may be written as a sum of $2N$ Bloch waves

$$\psi(r) = \sum_{j=1}^{2N} \alpha^{(j)} b^{(j)}(k^{(j)}, r) = \sum_{j=1}^{2N} \alpha^{(j)} \sum_g C_g^{(j)} \exp[i(k_0 + \gamma^{(j)} n + g) \cdot r]$$

in which $\alpha^{(j)}$ is the excitation amplitude of the jth Bloch wave. It should be noted, however, that to the accuracy of a phase factor, the Bloch wave solutions of equation (9.11) are periodic in reciprocal space, i.e.

$$b^{(j)}(k^{(j)}, r) = b^{(j)}(k^{(j)} + g, r). \tag{9.12}$$

It is a corollary of this periodicity that not all Bloch waves resulting from equation (9.11) are separate and distinct. Those which differ in k by only a reciprocal lattice vector are physically equivalent [20, 21].

For RHEED involving a total of n rods and N points of reciprocal lattice, there exist a total of $2n$ independent Bloch waves, and N strongly excited diffracted beams within the crystal. The total electron wave field within the crystal slab can be written as

$$\psi(r) = \sum_{j=1}^{2n} \alpha^{(j)} \sum_{g=g_1}^{g_N} C_g^{(j)} \exp[i(k_0 + \gamma^{(j)}n + g) \cdot r] \tag{9.13}$$

and the diffracted beam amplitude associated with the mth reciprocal lattice rod is given by

$$\psi_m(z) = \sum_{j=1}^{2n} \alpha^{(j)} \sum_{g_{m_z}} C_{m,g_{m_z}}^{(j)} \exp[i(k_{0_z} + \gamma^{(j)} + g_{m_z})z]. \tag{9.14}$$

Here we have used the notation $g = (m, g_{m_z})$ such that m denotes the two-dimensional reciprocal lattice vector parallel to the surface of a rod and g_{m_z} denotes the component along the rod.

The boundary conditions are given by continuation conditions imposed on the diffracted beam amplitudes and their surface normal derivatives. Neglecting a constant phase factor we have

$$\psi_m(z) = \sum_{j=1}^{2n} \alpha^{(j)} \sum_{g_{m_z}} C_{m,g_{m_z}}^{(j)} \exp[i(\gamma^{(j)} + g_{m_z})z] \tag{9.15}$$

$$-i\psi_m'(z) = \sum_{j=1}^{2n} \alpha^{(j)} \sum_{g_{m_z}} (k_{0_z} + \gamma^{(j)} + g_{m_z}) C_{m,g_{m_z}}^{(j)} \exp[i(\gamma^{(j)} + g_{m_z})z] \tag{9.16}$$

where the prime attached to ψ_m denotes differentiation with respect to z. Equations (9.15) and (9.16) can be written in matrix notation [22]

$$\boldsymbol{\Psi}(z) = \mathbf{S}\mathbf{P}(z)\mathbf{C}\boldsymbol{\Upsilon}(z)\boldsymbol{\alpha} \tag{9.17}$$

where $\boldsymbol{\Psi}$ is a $2n$-dimensional super-vector

$$\boldsymbol{\Psi}(z) = \begin{pmatrix} \{\psi_m(z)\} \\ \{-i\psi_m'(z)\} \end{pmatrix} \tag{9.18}$$

$\mathbf{S}$ is a $(2n \times 2N)$ matrix whose elements are

$$\{\mathbf{S}\}_{mg} = \begin{cases} 1 & \text{if } g \text{ belongs to } m\text{th rod} \\ 0 & \text{otherwise} \end{cases} \tag{9.19}$$

for $m \leqslant n$, $g \leqslant N$, and

$$\{\mathbf{S}\}_{m+n,g+N} = \{\mathbf{S}\}_{m,g}; \quad \{\mathbf{S}\}_{m+n,g} = \{\mathbf{S}\}_{m,g+N} = 0$$

$\mathbf{P}$ is a $(2N \times 2N)$ diagonal matrix:

$$\{\mathbf{P}\}_{gh} = \exp(\mathrm{i}g_z z)\delta_{gh} \tag{9.20}$$

for $h, g \leqslant N$, $g = (m, p)$, and

$$\{\mathbf{P}\}_{g+N,h+N} = \{\mathbf{P}\}_{g,h} \tag{9.21}$$

$\mathbf{C}$ is a $(2N \times 2n)$ matrix:
$$\{\mathbf{C}\}_{hi} = C_h^{(i)} \tag{9.22}$$

for $h \leqslant N$, and
$$\{\mathbf{C}\}_{h+N,i} = (k_{0_z} + \gamma^{(i)} + h_z)C_h^{(i)}$$

Υ is a $(2n \times 2n)$ diagonal matrix:

$$\{\Upsilon\}_{ij} = \exp(\mathrm{i}\gamma^{(i)}z)\delta_{ij} \tag{9.23}$$

and $\boldsymbol{\alpha}$ is a $2n$-dimensional column vector $(\boldsymbol{\alpha})_j = \alpha^{(j)}$.

If z_t and z_b are the coordinates of the top and bottom surfaces of a crystal slab, then from equation (9.17) we have

$$\Psi(z_t) = \mathbf{SP}(z_t)\mathbf{C}\Upsilon(z_t)\boldsymbol{\alpha}$$

which gives
$$\boldsymbol{\alpha} = [\mathbf{SP}(z_t)\mathbf{C}]^{-1}\Upsilon(-z_t)\Psi(z_t).$$

Substitution of the above equation into equation (9.17) gives an expression relating the vectors Ψ at the top and the bottom surfaces of the crystal slab

$$\Psi(z_b) = \mathbf{M}(z_b, z_t)\Psi(z_t) \tag{9.24}$$

in which the matrix $\mathbf{M}$ is called the *transfer matrix* of the slab and is given by

$$\mathbf{M}(z_b, z_t) = [\mathbf{SP}(z_b)\mathbf{C}]\Upsilon(t)[\mathbf{SP}(z_t)\mathbf{C}]^{-1}$$

in which $t = z_b - z_t$ is the thickness of the crystal slab.

9.3   SLICE METHODS—TWO-DIMENSIONAL BLOCH WAVES

9.3.1   Basic equations

Roughly speaking, the solution of an N-beam eigenvalue equation (9.3) scales as $\mathcal{O}[(n \times n_z)^3]$ in which n is the number of reciprocal lattice rods and n_z is the averaged number of reciprocal lattice points per rod along the surface normal direction. Typically n_z needs to be of the order of 10, demanding 1000 times more computing time than a system involving only two-dimensional varying potential, i.e. a z-independent potential field. Numerically a more efficient algorithm consists in using two-dimensional Bloch waves. In this scheme the potential variation along the surface normal direction is simulated by dividing the surface region and the bulk repeat unit slab into many slices, and assuming that the two-dimensional potential field parallel to the surface is constant normal to the surface (see figure 9.3). For a typical metal, such as a single crystal of silver, accurate results may be obtained by cutting a unit cell of about 0.4 nm into 100 slices. The scheme is then expected to be many times faster than the three-dimensional Bloch wave method. Other advantages of using two-dimensional Bloch waves include the reduction of very large asymmetric matrices of the order of tens of thousands into smaller and nearly symmetric matrices of the order of 100. Numerically speaking, results obtained by solving eigensystems of symmetric and smaller matrices are therefore much more reliable and accurate than solving the three-dimensional eigenvalue equations involving much larger and asymmetric matrices. Last but not the least, by employing two-dimensional Bloch waves one is able to deal with potential variation of any form along the surface normal direction, and therefore solve the fundamental difficulty encountered by using the original Bethe theory as discussed in the previous section.

We start our discussions of the two-dimensional Bloch wave approach by considering the basic equation (9.3) for one thin slice. Since the potential field in each thin slice is constant normal to the surface, we have for all reflections appearing in equation (9.3) $g = (G, g_z) = G$, giving for a single slice

$$\{(\gamma + k_{0_z})^2 - [K^2 - (k_0 + G)_t^2]\}C_G - \sum_{H \neq G} U_{GH} C_H = 0$$

or

$$(\mathcal{K}^2 - \Gamma_G^2)C_G - \sum_{H \neq G} U_{GH} C_H = 0 \qquad (9.25)$$

in which $\mathcal{K} = \gamma + k_{0_z}$, $\Gamma_G^2 = K^2 - (k_0 + G)_t^2$, and the subscript t denotes the tangential component of the vector parallel to the surface. The eigensystem (9.25) is indeed a first-order eigensystem in $\mathcal{K}^2$, giving $2n$ distinct eigenvalues

$$\gamma^{(i)} = \mathcal{K}^{(1)}, \mathcal{K}^{(2)}, \ldots, \mathcal{K}^{(n)}, -\mathcal{K}^{(1)}, -\mathcal{K}^{(2)}, \ldots, -\mathcal{K}^{(n)} \quad (i = 1, \ldots, 2n)$$

leading to $2n$ independent Bloch waves and n independent eigenvectors:

$$\Phi^{(i)} = (C_1^{(i)}, C_2^{(i)}, \ldots, C_n^{(i)}) \quad (i = 1, \ldots, n).$$

In general for an n-rod RHEED case the eigensystem (9.25) always gives n forward propagating and n backward propagating Bloch waves, and these Bloch waves as well as their wave vectors $\mathcal{K}^{(i)}$ are symmetric with respect to a mirror inversion of each crystal slice, and this is a natural consequence of the assumption that within each thin crystal slice the potential field is constant along the surface normal direction.

The diffracted beam amplitude associated with the mth reciprocal lattice rod (9.15) and its surface normal derivative (9.16) reduces to

$$\Psi(z) = \begin{pmatrix} \{\psi_m(z)\} \\ \{-i\psi'_m(z)\} \end{pmatrix} = \mathbf{C}\Upsilon(z)\alpha \tag{9.26}$$

in which $\Upsilon(t)$ is a $2n \times 2n$ diagonal matrix with

$$\{\Upsilon(t)\}_{i,i} = \exp(i\mathcal{K}^{(i)}t) \qquad \Upsilon(t)\}_{i+n,i+n} = \exp(-i\mathcal{K}^{(i)}t) \qquad (i = 1, \ldots, n)$$

and

$$\mathbf{C} = \begin{pmatrix} \{C_g^{(i)}\} & \{C_g^{(i)}\} \\ \{\mathcal{K}^{(i)}C_g^{(i)}\} & \{-\mathcal{K}^{(i)}C_g^{(i)}\} \end{pmatrix}.$$

For a thin slice of thickness t, we have

$$\Psi(z + t) = \mathbf{M}(t)\Psi(z) \tag{9.27}$$

with

$$\mathbf{M}(z) = \mathbf{C}\Upsilon(t)\mathbf{C}^{-1}.$$

By writing the inverse matrix of $\{C_G^{(i)}\}$ as $\{\overline{C}_G^{(i)}\}$ the inverse matrix of $\mathbf{C}$ is then given by

$$\mathbf{C}^{-1} = \frac{1}{2}\begin{pmatrix} \{\overline{C}_g^{(i)}\} & \{\overline{C}_g^{(i)}/\mathcal{K}^{(i)}\} \\ \{\overline{C}_g^{(i)}\} & \{-\overline{C}_g^{(i)}/\mathcal{K}^{(i)}\} \end{pmatrix}$$

and the transfer matrix $\mathbf{M}$ for a slice of thickness t is given by

$$\mathbf{M}(t) = \mathbf{C}\Upsilon(t)\mathbf{C}^{-1}$$
$$= \begin{pmatrix} \{\sum_i C_G^{(i)} \cos(\mathcal{K}^{(i)}t)\overline{C}_H^{(i)}\} & \{i\sum_i C_G^{(i)} \sin(\mathcal{K}^{(i)}t)\overline{C}_H^{(i)}/\mathcal{K}^{(i)}\} \\ \{i\sum_i \mathcal{K}^{(i)}C_G^{(i)} \sin(\mathcal{K}^{(i)}t)\overline{C}_H^{(i)}\} & \{\sum_i C_G^{(i)} \cos(\mathcal{K}^{(i)}t)\overline{C}_H^{(i)}\} \end{pmatrix}.$$

The inverse of the transfer matrix $\mathbf{M}(t)$ may be obtained simply by inverting the direction of propagation, i.e. $\mathbf{M}^{-1}(t) = \mathbf{M}(-t)$.

9.3.2   RHEED from the surface of a crystal slab

We will now consider the calculation of dynamical diffracted beam amplitudes from the surface of a slab system [23–25]. If the thickness of the crystal slab is much larger than the length of absorption, the slab method results effectively in the same reflection coefficients as from the surface of a semi-infinite crystal. In principle the transfer matrix could be constructed for the whole slab using (9.27). In practice, however, the transfer matrix as defined in (9.27) contains some exponential terms of the form $\exp(i\mathcal{K}^{(j)}z)$. For evanescent waves with negative imaginary $\mathcal{K}^{(j)}$ the transfer matrix $\mathbf{M}$ will diverge rapidly as the thickness of the crystal slab increases. Numerically, a better means for calculating RHEED from an assembly of slices is to propagate an $\mathbf{R}$ matrix which is defined as the ratio between the surface normal derivative of the wave function vector and itself [24, 26]. Assuming our model system consists of a total of m repeating bulk unit slabs and the surface region, then starting from the bottom surface of the bulk slab, we have

$$\begin{pmatrix} \{\psi_m\} \\ \{\psi'_m\} \end{pmatrix}_{m-1} = \begin{pmatrix} \mathbf{M}_{11} & \mathbf{M}_{12} \\ \mathbf{M}_{21} & \mathbf{M}_{22} \end{pmatrix}_m^{-1} \begin{pmatrix} \{\mathcal{T}_n \exp(ik_{nz}t)\} \\ \{k_{n_z}\mathcal{T}_n \exp(ik_{nz}t)\} \end{pmatrix}$$

$$= \begin{pmatrix} \{\overline{M}'_{11} + \overline{M}'_{12}k_{n_z}\}_{mn} \\ \{\overline{M}'_{21} + \overline{M}'_{22}k_{n_z}\}_{mn} \end{pmatrix}_m \{\mathcal{T}_n \exp(ik_{nz}t)\}_m$$

where $\mathcal{T}_n$ is the transmitted beam amplitude associated with the nth reciprocol lattice rod, and

$$\overline{\mathbf{M}}_m(t) = \mathbf{M}_m^{-1}(t) = \mathbf{M}_m(-t).$$

Letting

$$\{\mathbf{R}\}_k = \{M'_{21} + M'_{22}k_{n_z}\}\{M'_{11} + M'_{12}k_{n_z}\}^{-1} \tag{9.28}$$

we have then

$$\{\mathbf{\Psi}'\}_{m-1} = \{\mathbf{R}\}_m\{\mathbf{\Psi}\}_{m-1} \tag{9.29}$$

where $\mathbf{R}_m$ is the $\mathbf{R}$ matrix of the mth bulk repeat unit slab.

The $\mathbf{R}$ matrix can be propagated upward through the remaining bulk crystal slab composed of repeating unit slabs with indices $i = 1,\ldots,m-1$. Defining the $\mathbf{R}$ matrix general as in equation (9.29), we have for the ith slice ($i = k-1,\ldots,1$):

$$\begin{pmatrix} \{\psi_m\} \\ \{\psi'_m\} \end{pmatrix}_{i-1} = \begin{pmatrix} \overline{\mathbf{M}}_{11} & \overline{\mathbf{M}}_{12} \\ \overline{\mathbf{M}}_{21} & \overline{\mathbf{M}}_{22} \end{pmatrix}_i \begin{pmatrix} \{\psi_n\} \\ \{\psi'_n\} \end{pmatrix}_i$$

$$= \begin{pmatrix} \overline{\mathbf{M}}_{11} & \overline{\mathbf{M}}_{12} \\ \overline{\mathbf{M}}_{21} & \overline{\mathbf{M}}_{22} \end{pmatrix}_i \begin{pmatrix} \{\delta_{nk}\} \\ \{R_{nk}\} \end{pmatrix}_i \{\psi_k\}_i$$

giving

$$\{\mathbf{\Psi}\}'_{i-1} = \{\mathbf{R}\}_{i-1}\{\mathbf{\Psi}\}_{i-1} \tag{9.30}$$

where

$$\{\mathbf{R}\}_{i-1} = \{\overline{\mathbf{M}}_{21} + \overline{\mathbf{M}}_{22}\mathbf{R}_i\}\{\overline{\mathbf{M}}_{11} + \overline{\mathbf{M}}_{12}\mathbf{R}_i\}^{-1}. \tag{9.31}$$

At the surface region/substrate interface, we have then an equation for the reflected beam amplitudes, $\mathcal{R}_m$:

$$\begin{pmatrix} \{\delta_{m1} + \mathcal{R}_m\} \\ \{k_{m_z}(\delta_{m1} - \mathcal{R}_m)\} \end{pmatrix} = \begin{pmatrix} \{\psi_m\} \\ \{\psi_m'\} \end{pmatrix}_0 = \begin{pmatrix} \{\delta_{mn}\} \\ \{R_{mn}\}_0 \end{pmatrix} \{\psi_n\}_0 \tag{9.32}$$

and therefore

$$\{\mathcal{R}\}_m = -\{\delta_{mn} + R_{mn}^{-1}k_{n_z}\}^{-1}\{\delta_{mn} - R_{mn}^{-1}k_{n_z}\}\delta_{n1}. \tag{9.33}$$

When evanescent waves with large imaginary components of $\gamma^{(i)}$ are present, the transfer matrix for even a single repeat bulk unit slab or the surface region may diverge. In this case, both the surface region and the bulk unit slab need to be divided into several subslabs and the propagation of the $\mathbf{R}$ matrix through these subslabs follows exactly the same procedure as described as above.

9.3.3 RHEED from the surface of a semi-infinite crystal

We now consider the calculation of RHEED beam intensities from the surface of a semi-infinite crystal using the two-dimensional Bloch wave method [27]. On the one hand in a periodic bulk crystal we have

$$\mathbf{\Psi}(z + c) = \mathbf{M}(c)\mathbf{\Psi}(z) \tag{9.34}$$

where c is the lattice constant along the surface normal direction and the matrix $\mathbf{M}(c)$ appearing in (9.34) is the transfer matrix associated with a bulk repeat unit slab.

For a Bloch wave $b(\mathbf{r})$ in a crystal, it can be readily shown that both the Gth component b_G and its surface normal derivative $b_G' = \mathrm{d}(b_G)/\mathrm{d}z$ satisfy the Bloch theorem

$$b_G^{(j)}(z + c) = \exp(\mathrm{i}\gamma^{(j)}c)b_G^{(j)}(z) \qquad b_G'^{(j)}(z + c) = \exp(\mathrm{i}\gamma^{(j)}c)b_G'^{(j)}(z)$$

and that they are therefore both Bloch waves. We define a super-vector $b^{(j)}$ by

$$b^{(j)} = \begin{pmatrix} \{b_G^{(j)}\} \\ \{-\mathrm{i}b_G^{(j)}\} \end{pmatrix}. \tag{9.35}$$

Since equation (9.34) holds for a general wave function it also holds for $b^{(j)}$ defined by (9.35), i.e.

$$b^{(j)}(z + c) = \exp(\mathrm{i}\gamma^{(j)}c)b^{(j)}(z) = \mathbf{M}(c)b^{(j)}(z) \tag{9.36}$$

both $\gamma^{(j)}$ and Bloch waves $b^{(j)}$ may be obtained by diagonalizing the transfer matrix $\mathbf{M}(c)$ of a repeat bulk unit slab.

For a dynamical RHEED calculation involving n reciprocal lattice rods, the transfer matrix $\mathbf{M}$ is a $2n \times 2n$ matrix. In general this $2n \times 2n$ matrix will give a total of $2n$ Bloch waves, of which n propagate downwards into the crystal slab, and in the presence of absorption decay in amplitude downwards. The other n propagate and decay in the reverse direction. By introducing a $2n \times 2n$ matrix $\mathbf{C} = \{b^{(1)}, \ldots, b^{(2n)}\}$ and a diagonal matrix $\mathbf{\Upsilon}(z) = \{\exp(i\gamma^{(j)}z)\}$ we then have

$$\mathbf{M}(z) = \mathbf{C}\mathbf{\Upsilon}(z)\mathbf{C}^{-1}. \tag{9.37}$$

For a crystal slab consisting of m repeating unit slabs, substitution of equation (9.37) into (9.34) gives

$$\mathbf{\Psi}(z + mc) = \mathbf{C}\mathbf{\Upsilon}(mc)\mathbf{C}^{-1}\mathbf{\Psi}(z). \tag{9.38}$$

Without loss of generality we can assume that among the total $2n$ Bloch waves the first n Bloch waves are evanescent waves and the remaining waves are anti-evanescent. For a semi-infinite crystal these anti-evanescent Bloch waves are not physically allowed and must be discarded. Assuming that the interface between the surface region and the bulk crystal is at $z = z_s$ and that the transfer matrix associated with the surface region is $\mathbf{M}_s$, we then have for a crystal slab consisting of m repeating unit slabs

$$\mathbf{\Psi}(z + mc) = \mathbf{C}\mathbf{\Upsilon}(mc)\mathbf{C}^{-1}\mathbf{M}_s\mathbf{\Psi}(0)$$

$$= \mathbf{C}\begin{pmatrix} \{\exp(i\gamma^{(i)}mc)\} & \\ & \{\exp(i\gamma^{(i+n)}mc)\} \end{pmatrix}\mathbf{C}^{-1}$$

$$\times \mathbf{M}_s\begin{pmatrix} \{\delta_{G0} + \mathcal{R}_G\} \\ \{k_{G_z}(\delta_{G0} - \mathcal{R}_G)\} \end{pmatrix} \tag{9.39}$$

in which $\mathcal{R}_G$ is the reflected beam amplitude associated with the Gth reciprocal lattice rod. For a semi-infinite crystal, since all the physically allowed quantities must have a finite amplitude, the lower half of the column vector $\mathbf{C}^{-1}\mathbf{M}_s\mathbf{\Psi}(0)$ on the right-hand side of equation (9.39) must vanish. In terms of the two lower submatrices $\mathbf{M}'_{21}$ and $\mathbf{M}'_{22}$ of the matrix $\mathbf{M}' = \mathbf{C}^{-1}\mathbf{M}_s$ and the surface reflected beam amplitude vector $\{\mathcal{R}_G\}$, we can write the condition as

$$M'_{21}\{\delta_{G0} + \mathcal{R}_G\} + M'_{22}\{k_{G_z}(\delta_{G0} - \mathcal{R}_G)\} = 0 \tag{9.40}$$

which gives

$$\{R\}_G = -\left(\{M'_{21} - M'_{22}k_{H_z}\}_{GH}\right)^{-1}\left(\{M'_{21} + M'_{22}k_{0_z}\}_{H0}\right). \tag{9.41}$$

For a surface region with a simple structure, the transfer matrix $\mathbf{M}_s$ associated with the surface region can be simple and convergent if the transfer matrix associated with the bulk slab $\mathbf{M}(c)$ is convergent. However, for a complicated surface structure, the thickness of the surface region could be larger than that of a bulk slab and its transfer matrix $\mathbf{M}_s$ could be divergent even if that associated with the bulk slab is convergent. In this case we can propagate the $\mathbf{R}$ matrix rather than the transfer matrix through the surface region [27].

Alternatively the Bloch waves within the bulk crystal may be obtained by taking advantage of the fact that within the bulk crystal only half of the $2n$ Bloch waves are physically allowed and the remaining half must have zero excitation amplitudes, i.e. for these Bloch waves $\alpha^{(i)} = 0$. Within the bulk crystal we therefore have

$$\psi_G(z) = \sum_{i=1}^{n} \alpha^{(i)} C_G^{(i)} \exp(i\mathcal{K}z) \tag{9.42}$$

or in matrix notation

$$\{\psi_G(z)\} = \mathbf{c}\boldsymbol{\gamma}\boldsymbol{\alpha}$$

in which $\mathbf{c} = \{C_G^{(i)}\}$ is an $n \times n$ matrix, $\boldsymbol{\gamma}$ is a diagonal matrix with $\{\gamma\}_i = \exp(i\mathcal{K}^{(i)}z)$ and $\boldsymbol{\alpha}$ is an n-dimensional vector with $\{\alpha\}_i = \alpha^{(i)}$. For a crystal thickness of $t + d$, d being the lattice constant along the surface normal direction, we have

$$\{\psi_G(z + d)\} = \mathbf{c}\boldsymbol{\gamma}(d)\mathbf{c}^{-1}\{\psi_G(z)\}. \tag{9.43}$$

On the other hand, from equation (9.34) we have

$$\{\psi_G(z + d)\} = \mathcal{M}\{\psi_G(z)\} \tag{9.44}$$

with

$$\mathcal{M} = \prod_k \left(\mathbf{M}_{11}(t_k) + \mathbf{M}_{12}(t_k)\mathbf{R}_k \right)_{k-1}$$

in which $\mathbf{R}_k$ is the $\mathbf{R}$ matrix of the kth thin slice as defined by equation (9.30), k runs over all thin slices composing the repeat unit slab and $d = \sum_k t_k$. Comparing equations (9.43) and (9.44) we see that the eigenvectors $\{C_G^{(i)}\}$ and eigenvalues $\{\exp(i\mathcal{K}^{(i)}d)\}$ of the Bloch waves within the bulk may be obtained directly by the diagonalization of the matrix $\mathcal{M}$, and this is equivalent to the method given by Dudarev [28].

9.4 NUMERICAL EXAMPLES

9.4.1 Structural model and scattering potential

In this subsection we will consider the case of an Ag(001) surface as an example to illustrate the procedure of constructing the scattering potential. A silver single

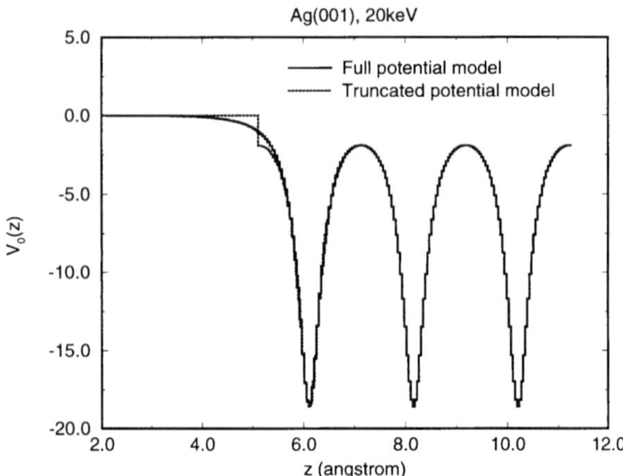

Figure 9.3 Real and imaginary parts of the averaged optical potential $V_0(z)$ at the near surface region containing the surface region and one repeat bulk unit slab. The potential distribution is calculated for a silver single crystal and is approximated by a step function. The z direction is along $[00\bar{1}]$ and the incident beam energy is 20 keV.

crystal has the f.c.c. structure with lattice constant $a_0 = 4.09$ Å and four atoms within the conventional unit cell. The fractional atomic coordinates for the four atoms are $(0, 0, 0)$, $(1/2, 1/2, 0)$, $(1/2, 0, 1/2)$ and $(0, 1/2, 1/2)$. In what follows we use an indexing system calling all reflections (H, K) with non-zero negative K value higher-order Laue zone (HOLZ) reflections and all reflections with non-zero positive K value minus HOLZ (MHOLZ) reflections. All reflections having indices of the form $(H,0)$ are called zero-order Laue zone (ZOLZ) reflections. The high energy electrons are incident on the (001) surface of the silver single crystal along the [01] direction, which corresponds to the conventional [110] azimuth.

To simulate the potential variation from the substrate to the vacuum, a large unit cell of the following lattice parameters is constructed: $a = a_0/\sqrt{2} = 2.892$ Å (along the conventional $[\bar{1}10]$ direction), $b = a_0/\sqrt{2} = 2.892$ Å (along the conventional [110] direction), and $c = 4a_0 = 16.36$ Å (along the conventional [001] direction). The z axis is chosen to point toward the conventional $[00\bar{1}]$ direction, the beam azimuth is along the [01] direction, and the primary beam energy used is 20 keV. Six silver atoms are placed in the large unit cell, with fractional atomic coordinates $(0, 0, 6/16)$, $(1/2, 1/2, 8/16)$, $(0, 0, 10/16)$, $(1/2, 1/2, 12/16)$, $(0, 0, 14/16)$ and $(1/2, 1/2, 16/16)$. The surface region is taken to be the slab in between $z = 0$ and $z = 7.1575$, and the repeat unit slab to be that in between $z = 7.1575$ and $z = 11.2475$. Figure 9.3 shows the averaged one-dimensional potential

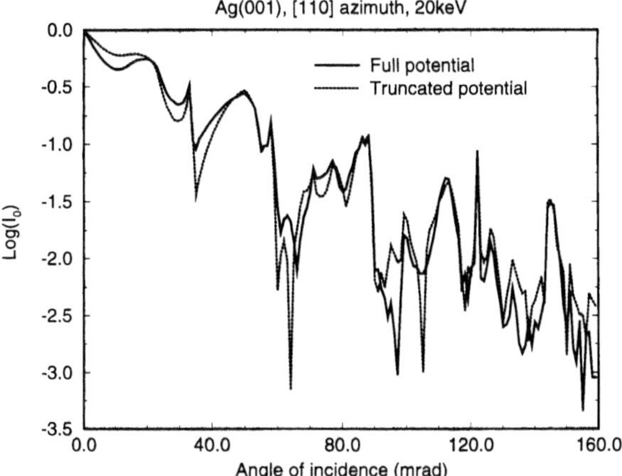

Figure 9.4 Specular reflected beam intensities versus angle of incidence for (- - - -) a truncated potential model and (———) a full potential model. The calculations were made for a primary beam energy of 20 keV and an Ag(001) surface. The high energy electrons are incident on the surface along the [110] zone axis.

distributions $V_0(z)$ through the surface region and one repeat bulk unit slab for (a) a truncated potential model as originally used by Bethe and (b) a full potential model with smooth transition from the interior of the crystal into the vacuum. It should be noted that in figure 9.3 the continuously varying potential has been replaced by a stepped function in accordance with the assumption from which the two-dimensional Bloch wave formula is derived.

Figure 9.4 shows two curves of specular beam intensities versus the angle of incidence for a truncated potential model (dotted line) and a full potential model (solid line). While there exists a certain similarity between the two curves, in particular for peak positions, the influence of the gradual transition between the potential field exterior to and in the interior of a crystal is obviously not negligible for a wide range of angles of incidence. Qualitatively this result may be attributed to the greatly reduced electron wave penetration depth under certain diffraction conditions, especially the surface resonance conditions. Under these conditions the electron penetration depth may be as small as a monatomic layer, and an artificial step function introduced by the truncated potential model at the crystal/vacuum interface is expected to produce pronounced effects as shown in figure 9.4. It may therefore be concluded that the influence of the gradual transition between the potential field exterior to and in the interior of a crystal is in general important in RHEED diffraction geometry even for a primary energy of as high as several tens of kilovolts.

The optical potential appearing in the eigensystems (9.3) and (9.25) can be calculated most conveniently by representing atomic scattering factors as a sum of Gaussians [29, 5, 26]

$$f^{(e)}(s) = \sum_j a_j \exp(-b_j s^2) \tag{9.45}$$

where $s = 2\pi k \sin\theta = \sin\theta/\lambda$, θ being the half angle of scattering and $\lambda = 2\pi/k$ being the electron wavelength, and a_j and b_j are fitting parameters. For the case of neutral atoms these parameters are now known for all the elements of the periodic table [30] and 106 ions [31]. Assuming that the potential of interaction between the incident electron and the crystal can be approximated by the sum of potentials of individual atoms or ions, and using the mixed real and reciprocal space representation [32, 33] we obtain

$$V(\mathbf{R}, z) = \sum_G V_G(z) \exp(i\mathbf{G}\cdot\mathbf{R}) \tag{9.46}$$

where $\mathbf{G}$ denotes a two-dimensional reciprocal lattice vector parallel to the surface and $V_G(z)$ is given by

$$V_G(z) = \frac{4\pi\hbar^2}{S_0 m_0} \sum_{n,i} \exp(-i\mathbf{G}\cdot\mathbf{R}_n)$$

$$\times \left\{ a_{i,n}^{(Re)} \sqrt{\frac{\pi}{b_{i,n}^{(Re)} + B_n}} \exp\left[-\frac{(b_{i,n}^{(Re)} + B_n)G^2}{(4\pi)^2} - \frac{4\pi^2}{b_{i,n}^{(Re)} + B_n} \right] \right.$$

$$\left. + ia_{i,n}^{(Im)} \sqrt{\frac{\pi}{b_{i,n}^{(Im)} + B_n/2}} \exp\left[-\frac{(b_{i,n}^{(Im)} + B_n/2)G^2}{(4\pi)^2} - \frac{4\pi^2}{b_{i,n}^{(Im)} + B_n/2} \right] \right\} \tag{9.47}$$

in which m_0 is the rest mass of the electron, $\hbar$ is the Planck constant, the summation over n is carried out over all atoms belonging to a surface unit cell and S_0 denotes the area of the unit cell. It should be noted that the fitting parameters a_i and b_i of the absorptive atomic scattering factor are obtained by fitting the quantity $f^{(TDS)}(s)\exp(-Bs^2/2)$ rather than $f^{(TDS)}(s)$, i.e.

$$f^{(TDS)}(s)\exp(-Bs^2/2) = \sum_{j=1}^{5} a_j^{(Im)} \exp(-b_j^{(Im)} s^2).$$

This is because in the limit of large s the magnitude of the absorptive atomic scattering factor $f^{(TDS)}(s)$ increases exponentially with increasing s, i.e. $f^{(TDS)}(s) \sim \exp(Bs^2/2)$. For some elementary crystals and important materials with the zinc blende-structure the fitting parameters have been given in references [30, 34] for a range of temperature and a primary beam energy of 100 keV. These parameters may be converted for any beam energy E via the following relation

$$a_j(s, E) = [\beta(100 \text{ keV})/\beta(E)]a_j(s, 100 \text{ keV}) \qquad (9.48)$$

where $\beta = (1 - \gamma^{-2})^{1/2}$, $\gamma = (m/m_0) = 1.0 + 1.956\,9314 \times 10^{-3}E$, and the primary beam energy is in keV.

The amplitude of scattering of high energy electrons by an ion is very different from that of a neutral atom. For x-ray diffraction, the atomic scattering factors of both neutral atoms and ions satisfy the condition that

$$Z_0 = \lim_{s \to 0} f^{(X)}(s) \qquad (9.49)$$

where Z_0 is the number of electrons per atom which can either be in a neutral or in a charged ionic state. For a neutral atom $Z_0 = Z$, Z being the atomic number of the atom. For an ion $Z_0 \neq Z$, and the difference between the two quantities represents the excess or deficiency of charge on the nucleus resulting from charge transfer associated with the formation of chemical bonds in the crystal. The atomic scattering factor for electron diffraction is related to that for x-ray diffraction by the Mott formula [35]

$$f^{(e)}(s) = \frac{m_0 e^2}{8\pi^2 \hbar^2} \frac{Z - f^{(X)}(s)}{s^2} \qquad (9.50)$$

where e is the electron charge. For an ion where the number of electrons is not equal to the charge of the nucleus $Z \neq Z_0$, it follows from equation (9.50) that as s approaches zero the scattering factor diverges as $\sim (Z - Z_0)/s^2$.

Electron scattering factors of ions have been calculated numerically and tabulated by several authors including Doyle and Turner [29], Cowley [5] and Rez *et al* [36]. In principle *ab initio* numerical values may be fitted using the same analytical form (9.45) as for neutral atoms. However, since expression (9.45) *does not* diverge for the zero angle of scattering as it should do for an ion, it is not suitable for accurate fitting of ionic scattering factors [37].

The examination of the origin of the divergent behaviour of the electron scattering factor of an ion, see for example Doyle and Turner [29]), shows that the divergent part arises from the contribution of unscreened long range Coulomb potential. This may be readily demonstrated by rearranging equation (9.50) as

$$f^{(e)}(s) = \frac{m_0 e^2}{8\pi^2 \hbar^2} \frac{Z_0 - f^{(X)}(s)}{s^2} + \frac{m_0 e^2}{8\pi^2 \hbar^2} \frac{\Delta Z}{s^2} = f_0^{(e)}(s) + \frac{m_0 e^2}{8\pi^2 \hbar^2} \frac{\Delta Z}{s^2} \qquad (9.51)$$

where $\Delta Z = Z - Z_0$ represents the ionic charge and the second term on the right-hand side represents the Coulomb part of the scattering factor. The first term on the right-hand side (i.e. $f_0^{(e)}(s)$) results from scattering of electrons by the screened atomic field. The condition (9.49) ensures that $f_0^{(e)}(s)$ remains finite in the limit $s \to 0$.

On a purely numerical basis the crystal potential of an ionic crystal can be calculated using a three-dimensional super unit cell. The size of the super unit cell in the plane parallel to the surface equals the size of the surface unit cell, and it may be arbitrarily large in the direction normal to the surface. Using the super unit cell the potential can be calculated using conventional Fourier expansion both in the plane of the surface and in the direction normal to it. The Fourier components of the three-dimensional potential are given by [15]

$$V_g = -\frac{\hbar^2}{2m_0} \frac{4\pi}{\Omega} \sum_i f_i^{(e)}(s) \exp(i\boldsymbol{g} \cdot \boldsymbol{r}_i - B_i s^2) \tag{9.52}$$

where Ω is the volume of the super unit cell, $g = 4\pi s$ and $\boldsymbol{g} = (\boldsymbol{G}, \ell)$ is a three-dimensional reciprocal lattice vector and $\boldsymbol{r}_i$ is the coordinate of the ith atom. B_i is the Debye–Waller temperature factor which is assumed to be isotropic (for the general anisotropic case see [38]), and the summation is carried out over all atoms within the super unit cell. After all the components V_g have been calculated, the real space potential distribution can be obtained by using the inverse Fourier transform. For $V_G(z)$ given by equation (9.46) we obtain

$$V_G(z) = \sum_\ell V_{G,\ell} \exp(i\ell z). \tag{9.53}$$

As before, we separate $V_G(z)$ into two parts, namely the contribution due to ionic charge $\Delta V_G(z)$ and the remaining part associated with the screened atomic field $V_G^0(z)$. While the contribution resulting from $f_0^{(e)}(s)$ always remains finite, the part of the potential associated with the presence of ionic charge may give rise to divergent terms. For the termination of the crystal lattice satisfying the condition of charge neutrality and having zero total dipole moment, the averaged potential is given by

$$\Delta V_0(z) = \sum_{\ell \neq 0} \Delta V_\ell^0 \exp(i\ell z) + \frac{e^2}{4\pi\Omega} \sum_i \Delta Z_i (8\pi^2 z_i^2 + B_i). \tag{9.54}$$

Figure 9.5 shows the variation of the potential averaged in the (111) plane of UO_2. The surface of the crystal is assumed to be terminated by a layer of oxygen O^{2-} ions. Curves shown in the figure illustrate the distribution of the average potential calculated for the neutral atom model and ionic model of the crystal. In the latter case the contribution due to ionic charges and that due to the screened atomic field are shown as separate quantities. Although in this case z projections of negatively charged layers of O^{2-} ions and positively charged layers of U^{4+} ions do not coincide, the crystal as a whole still satisfies the condition of charge neutrality and has vanishing total dipole moment perpendicular to the surface. As

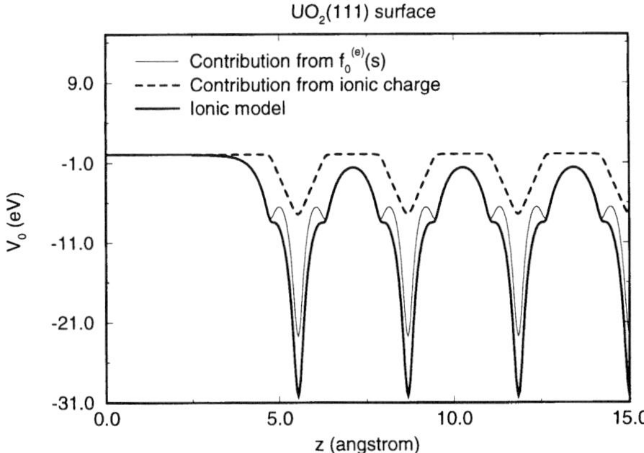

Figure 9.5 Averaged potential distribution of a single crystal of UO$_2$, with the averaging being carried out parallel to the (111) surface.

a result, contributions from positively and negatively charged planes cancel and the potential vanishes in the vacuum region above the surface. In a more general case where a crystal is characterized by an intermediate degree of ionicity α the electron scattering factor entering equation (9.52) may be expressed as

$$f^{(e)}(s) = (1 - \alpha) f^{(e)}_{\text{neutral}}(s) + \alpha f^{(e)}_{\text{ion}}(s) \tag{9.55}$$

where $f^{(e)}_{\text{neutral}}(s)$ and $f^{(e)}_{\text{ion}}(s)$ are scattering factors of neutral atoms and their ions, respectively. A recent study of RHEED from the (100) NiO surface shows that this scheme works well in describing electron diffraction by planes parallel to the surface [39].

9.4.2 Effects of ionic charges on the electrostatic potential

A simple analytical expression for $V_G(z)$ and zero thermal vibration, i.e. zero temperature factor $B = 0$, may be obtained following a procedure similar to that leading to equation (9.48). In equation (9.51) we represented the electron scattering factor of an ion by a sum of two terms where the second term is the contribution due to the ionic charge and the first term $f_0^{(e)}(s)$ is the contribution from the remaining screened atomic field. As shown above, the term $f_0^{(e)}(s)$ remains finite for all angles of scattering and it can therefore be fitted accurately by five Gaussians following equation (9.45). The contribution to $V_G(z)$ resulting from this term therefore takes the form which is identical to expression (9.48) and which vanishes in the vacuum region outside the crystal.

The contribution to the potential resulting from the second term of equation (9.51) describes effects associated with long-range Coulomb field giving rise to the real-space term of the form

$$\Delta\phi(\boldsymbol{r}) = \frac{e\Delta Z}{r}.$$
(9.56)

The corresponding contribution to $V_G(z)$ is

$$\Delta V_G(z) = -\frac{2\pi e^2}{S_0}\frac{1}{G}\sum_n \Delta Z_n \exp\left[-i\boldsymbol{G}\cdot\boldsymbol{R}_n - G|z - z_n|\right]$$
(9.57)

for $G \neq 0$ and

$$\Delta V_0(z) = \frac{2\pi e^2}{S_0}\sum_n \Delta Z_n|z - z_n|.$$
(9.58)

It should be pointed out that in deriving the above expression for $V_0(z)$ we used the condition of charge neutrality, i.e. $\sum_n \Delta Z_n = 0$. If we define the positive direction of the z-axis as pointing outwards then at a certain distance above the crystal surface for all z_n we have $z > z_n$ and

$$\Delta V_0(z) = -\frac{2\pi e^2}{S_0}\sum_n \Delta Z_n z_n.$$
(9.59)

In other words, the value of the potential in the vacuum is proportional to the z projection of the *total* dipole moment of the crystal $\sum_n \Delta Z_n z_n$. Depending on the geometry of termination of the crystal lattice we can identify three distinct classes of ionic surfaces. The first case corresponds to the situation where both positively and negatively charged ions are located in the same atomic plane, examples being the (001) and (110) terminations of the sodium chloride structure or the (110) plane of the fluorite structure (see figure 9.6(*b*)). For each plane $z_n = $ const and the z projection of the dipole moment vanishes, i.e.

$$\mu = \sum_n \Delta Z_n z_n = \text{const}\sum_n \Delta Z_n = 0.$$

In the vacuum region above the surface the potential rapidly goes to zero, and this case presents no particular difficulty to dynamical RHEED calculations.

In the second case positive and negative ions are located in different atomic planes so that each atomic plane parallel to the surface is charged, but the total dipole moment of the repeat unit still vanishes. Examples of this type of termination of the crystal lattice include the (111) surface of the fluorite structure, provided that the surface is terminated on the anion plane [40] (see figure 9.6(*c*)). In the fluorite structure atomic planes consist of neutral anion–cation–anion repeat units in the direction perpendicular to the (111) plane, such as an $O^{2-}-U^{4+}-O^{2-}$ unit in uranium dioxide or $F^{1-}-Ca^{2+}-F^{1-}$ unit in calcium fluoride. For each unit the total dipole moment is equal to zero and the scattering

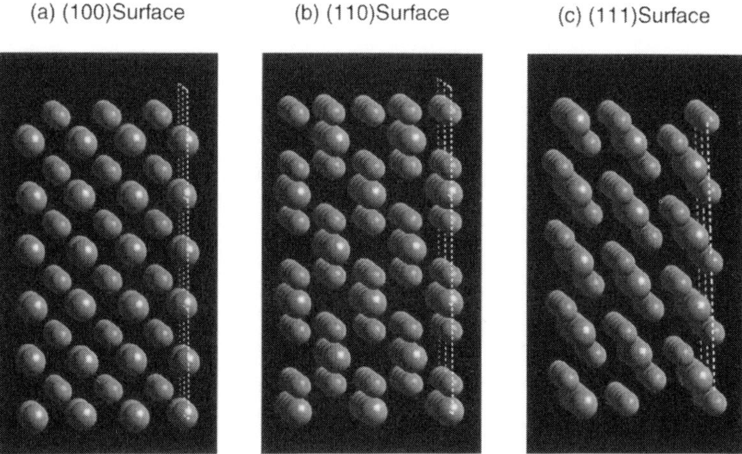

Figure 9.6 Atomic model for the (*a*) (100), (*b*) (110) and (*c*) (111) surfaces (shown by dotted lines) of the fluorite structure. In this figure the larger ion corresponds to the anion, e.g. U^{4+} ion in uranium dioxide, and the smaller ion corresponds to the cation, i.e. O^{2-}.

potential vanishes in the vacuum region. Dynamical RHEED calculation can again be performed using one of the conventional numerical techniques although care needs to be taken of the correct representation of the relevant Coulomb terms.

In the third case positive and negative ions do not lie in the same plane and the total dipole moment of a repeat unit of ionic planes is not equal to zero. Examples of this case include the (100) surface of the fluorite structure (see figure 9.6(*a*)) and the (111) surface in the sodium chloride structure. In uranium dioxide the repeat unit along the ⟨100⟩ direction consists of two ionic planes, i.e. $2O^{2-}$–U^{4+}, and for this repeat unit the total dipole moment has a finite value. As a result, the crystal potential diverges as the thickness of the crystal slab increases. Other physical quantities also show divergent behaviour (for example the surface energy of the (100) surface of UO_2 is infinite [40]) and this makes simple termination of the bulk structure impossible. This tendency can be formally eliminated by choosing a suitable surface reconstruction, for example in the case of the (100) surface of UO_2 this can be achieved by transferring one half of the top oxygen layer from one surface of the crystal slab to the other surface. In a real crystal the tendency towards lowering the surface energy also leads to surface reconstruction and/or to the generation of surface defects. One of the possible scenarios involves the formation of differently terminated steps exposing regions of oppositely charged surface planes [40]. Therefore, in any realistic situation the total

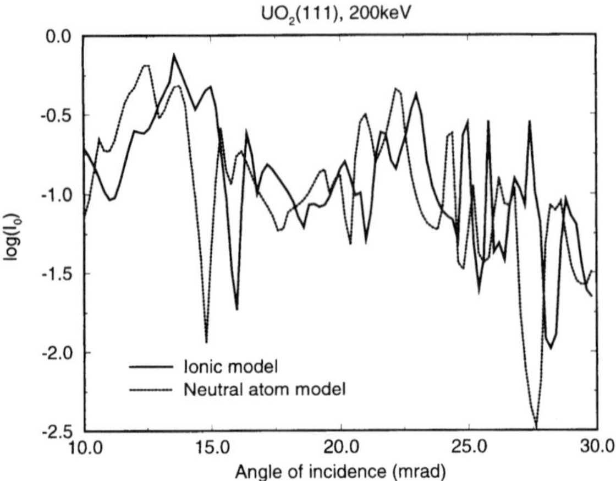

Figure 9.7 Calculated RHEED rocking curves for the specular reflected beam using a neutral atom model (dotted line) and an ionic model (solid line) for a single crystal of UO$_2$. The calculations were made for 200 keV primary beam energy, and a beam azimuth along [$\bar{1}\bar{1}2$]. The quantities plotted in the figure are the absolute amplitudes of the reflected beam amplitudes.

dipole moment of a repeat unit associated with a particular termination of the crystal lattice must vanish to eliminate the divergent behaviour of the potential in the limit of large thickness of the crystal slab. In should be noted that surface reconstruction may lead to the appearance of *surface* dipole moment and may therefore influence the apparent magnitude of the inner potential of the crystal seen in experiments on refraction of electrons by a crystal surface.

9.4.3 Effects of charge transfer on RHEED rocking curves

Once all the Fourier components of the potential $V_G(z)$ have been evaluated, dynamical RHEED calculations may be readily performed as discussed in the preceding section. Figure 9.7 shows a RHEED rocking curve calculated for the (111) surface of uranium dioxide UO$_2$. Electrons are incident on the surface in the direction of the [$\bar{1}\bar{1}2$] zone axis. Although for this surface negatively charged O^{2-} and positively charged U^{4+} planes are shifted with respect to each other in the direction normal to the surface, the z projection of the total dipole moment of each repeat unit of the form O^{2-}–U^{4+}–O^{2-} is equal to zero. It is seen that there exists a clear correlation between the rocking curves calculated using the neutral atom and the ionic models. The transfer of charge from U to oxygen ions results in a shift of rocking curves towards the region of higher angles, and this finding is consistent with what can be expected from the analysis of the distribution of crystal potential. The transfer of charge between ions leads mainly to the decrease of the inner potential. It has a less dramatic effect on the shape of the potential. Since the smaller value of the inner potential is equivalent to the reduction of the refractive index, the main effect of the charge transfer on RHEED rocking curves is therefore associated with the shift of all diffraction features towards the region of higher glancing angles as shown in figure 9.7.

It should also be pointed out that charge transfer in an ionic crystal does not only result in the homogeneous shift of intensity peaks. It also changes the relative peak heights and relative positions of peaks in RHEED rocking curves. Although in many cases there still exists a one-to-one correspondence between peaks in the RHEED rocking curves calculated using neutral atom and ionic models, charge transfer often leads to the appearance of new features and sometimes any correspondence between the two curves appears to be lacking. This shows the importance of using an adequate representation of electron scattering factors which would take into account the effects of charge transfer occurring in a real crystal. In a recent study of the NiO(001) surface [39] it was found that the mixed neutral atom/ionic model (9.55) gives rise to a reasonably good description of the one-rod RHEED case. It can be expected that in the case of many-rod RHEED scattering this scheme will also provide a starting point which is better than either the neutral atom model or the ionic model.

9.5 MBE GROWTH OF THIN FILMS AND RHEED INTENSITY OSCILLATIONS

9.5.1 Molecular beam epitaxy

Molecular beam epitaxy (MBE) is a versatile technique for growing multilayer structures made of semiconductors, metals or insulators. In MBE, thin films crystallize via reactions between incident molecular or atomic beams and a substrate surface which is maintained at an elevated temperature in ultrahigh vacuum, see for example [41]. What distinguishes MBE from other deposition techniques is its significantly more precise control of the beam fluxes and growth conditions by such surface diagnostic methods as Auger electron spectroscopy (AES) and RHEED. In particular the technique of RHEED intensity oscillations has been used routinely to calibrate beam fluxes and alloy compositions and to control the thicknesses of quantum wells and superlattice layers (see articles in Larsen and Dobson [4]).

9.5.2 RHEED intensity oscillations

The phenomenon of RHEED oscillations was first discovered by Harris, Joyce and Dobson [42] and Wood [43]. It was found that the intensities of all diffracted features oscillate with a period which under most circumstances corresponds

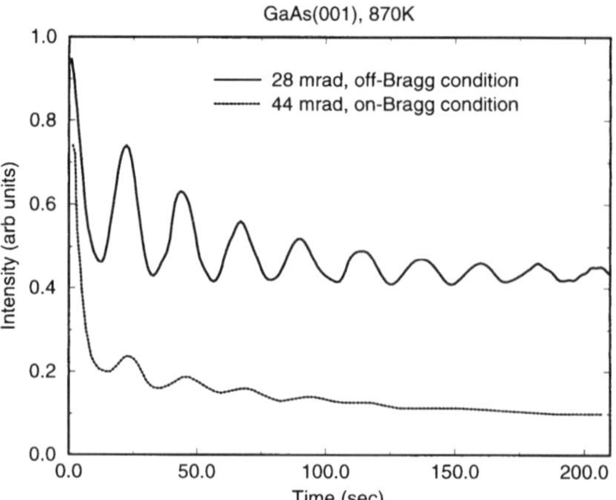

Figure 9.8   RHEED intensity oscillations for GaAs(001) surface and two angles of incidence, (——) 28 mrad corresponds to the off-Bragg condition and (- - - -) 44 mrad corresponds to the (400) Bragg condition. Experimental data are taken from Van Hove *et al* [64].

to the growth of a single atomic or molecular layer (see figure 9.8). This phenomenon was originally attributed to result from a periodically varying contribution of the diffusely scattered electrons [44] and the RHEED intensity was compared with surface step density and surface diffusion parameters were derived from the comparisons (see for example [45, 46]). The basic features of the surface step density model are illustrated in figure 9.9(*a*). The physical argument given in supporting the step density model is as follows. Since each surface step ledge acts as a localized source of diffuse scattering, an increase in the step density is therefore expected to cause an increase in diffuse scattering and a decrease in the specular reflected beam amplitude, i.e. the changes in the diffracted beam intensity might be equated with changes in surface roughness. A perfect defect-free smooth surface corresponds to the highest reflected beam intensity. When growth commences two-dimensional nuclei are formed at random positions on the crystal surface, the elastically reflected beam intensity decreases and diffusely scattered beam intensity increases. The elastic reflected beam intensity will be minimum at half-layer coverage and maximum at the start of each new layer.

9.5.3   Step density model

While the surface step density model is intuitively reasonable and has been widely accepted, much experimental evidence points to its inadequacy in describing RHEED intensity oscillations quantitatively and recently this model

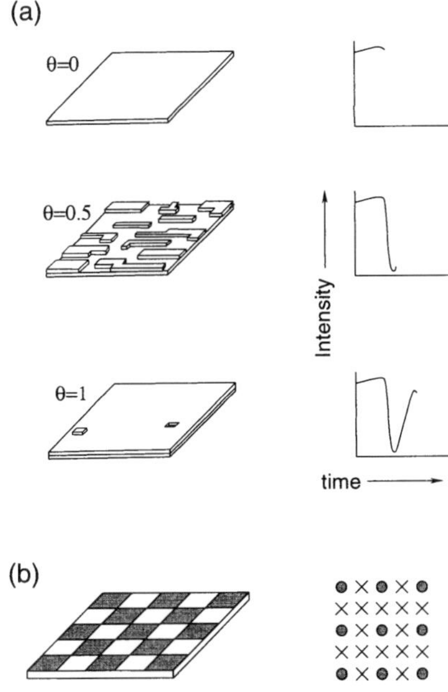

Figure 9.9 (*a*) A layer by layer growth model and expected RHEED intensity behaviour using a surface step density model. (*b*) A regular array of surface islands and corresponding electron diffraction pattern geometry.

has been explicitly criticized by Korte and Maksym [47]. Shown in figure 9.9(*b*) are regularly arranged surface islands. On the one hand, since there exist many surface islands on the surface with a half-layer coverage and high density of surface steps, a maximum diffuse scattering and a minimum elastic scattering would be expected in accordance with the surface step density model given above. One the other hand, noticing that the surface islands are occupying a regular array of sites on the surface, the general dynamical diffraction theory predicts that the integrated diffusely scattered electron beam intensity over reciprocal space in between Bragg spots will be zero for this particular island distribution, and this contradicts the prediction of the surface step density. The physical reason for this contradiction is that in the surface step density model the diffraction processes are assumed to be completely incoherent and diffusely scattered electron beams from surface step ledges add up in intensity resulting in strong diffuse scattering everywhere in between the Bragg spots. Actually in RHEED to an excellent approximation the diffraction processes occurring on the surface may be treated as coherent or at least partially coherent over many hundreds of surface islands. Diffusely scattered electron beams from the step ledges add up in amplitude rather than intensities, resulting in zero intensity everywhere except in discrete directions satisfying constructive interference or the Bragg conditions.

9.5.4 Kinematic approximation

An alternative explanation of RHEED intensity oscillation was put forward by Cohen and co-workers [48]. In its simplest form, a perfect layer by layer growth model is used and the specular reflected beam amplitudes from the two exposed surface layers f_1 and f_2 are assumed to be proportional to their respective reflecting areas, i.e. $f_1 \propto \theta$ and $f_1 \propto 1 - \theta$. Let $\mathbf{k}'$ and $\mathbf{k}$ be the incident and reflected electron wave vectors and h be the height of a monatomic surface step; we have then the following expression for the specular reflected beam intensity

$$I \propto |f_1 + f_2|^2 = |\theta + (1 - \theta) \exp[\mathrm{i}(\mathbf{k}' - \mathbf{k})_z h]|^2$$
$$= \theta^2 + (1 - \theta)^2 + 2\theta(1 - \theta) \cos[(\mathbf{k}' - \mathbf{k})_z h].$$

Two distinct diffraction conditions may be recognized. First the in-phase or Bragg condition when the following equation is satisfied: $(\mathbf{k}' - \mathbf{k})_z h = 2n\pi$, $n = 1, 2, \ldots$. For this condition the above expression predicts a constant reflected beam intensity for varying surface layer coverage θ, i.e. the diffracted beam intensity does not oscillate, and this is in contradiction with many experimental observations, see for example figure 9.8. Second the out-of-phase or off-Bragg condition when the equation $(\mathbf{k}' - \mathbf{k})_z h = (2n + 1)\pi$ is satisfied. For this diffraction condition the kinematic argument gives $I \propto 1 - 4\theta(1 - \theta)$, i.e. RHEED intensity oscillates periodically with varying layer coverage. For a perfect smooth surface, i.e. $\theta = 0$ or 1, the specular reflected beam intensity is a maximum and for a half-layer coverage, i.e. $\theta = 0.5$, the intensity is a minimum. While qualitatively predictions of this simple kinematic diffraction theory agree with some general features of experimentally observed RHEED intensity oscillations, e.g. RHEED intensity oscillations are more pronounced for off-Bragg conditions than for on-Bragg conditions (see figure 9.8), this theory fails in explaining many of the observed details of RHEED intensity oscillations. Among other features, experimentally observed intensity minima are often observed to correspond to an intensity which is intermediate between the intensity maximum and minimum [49], and this is the famous phase problem of RHEED intensity oscillations [50, 51]. In what follows we will show that these features result mainly from the multiple scattering effect and that most of the experimentally observed RHEED intensity oscillations' features may be reproduced employing a multiple scattering model of RHEED [52, 53, 54].

9.5.5 Simulations of MBE processes

MBE processes may be simulated by molecular dynamics or Monte Carlo methods (see for example [55, 45]), and rate equations [56, 57]. In the simplest

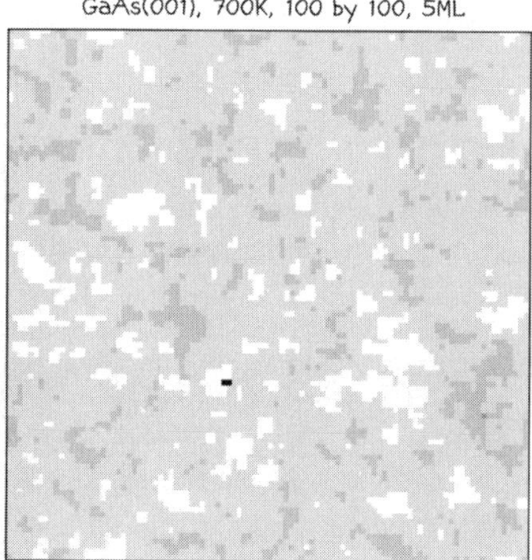

Figure 9.10 Snapshot of simulated surface morphologies for a GaAs(001) surface. In the picture 5.0 monolayers of atoms have been deposited upon a 100×100 lattice at 700 K.

Monte Carlo model the crystal is described by an array of columns of atoms, and the substrate is treated as a simple cubic lattice in which vacancies and overhangs are forbidden [58]. The growth is assumed to consists of two kinds of process: the deposition of a monatomic species onto the substrate, and the migration of surface adatoms. Evaporation of surface atoms is neglected [45]. Growth is initiated by the random deposition of atoms onto the surface at a rate FA, where F is the incident beam flux, and A is the area of the substrate. A site on the surface is selected randomly, and the most stable site having maximum number of nearest neighbours within a radius encompassing the nearest neighbour sites of the arrival point is chosen as the site of final deposition. Surface migration is modelled by an Arrhenius expression for the inter-site hopping rate $k(E, T) = k_0 \exp(-E/k_B T)$, where k_0 is the vibration frequency of a surface adatom, k_B is Boltzmann's constant, T is the substrate temperature and E is the local energy barrier to hopping. The energy barrier E consists of two terms, a substrate barrier E_s and a site-dependent term E_N which is proportional to the number of nearest neighbours, i.e. $E = E_s + n E_N$, where $n = 0, 1, 2, 3, 4$.

Shown in figure 9.10 is a snapshot of surface morphologies simulated using the simple Monte Carlo method for a GaAs(001) surface at 650 K. The simulations were made for a 100×100 lattice, a beam flux of $F = 1$ ML s^{-1}, a substrate energy barrier $E_s = 1.3$ eV and a site-dependent barrier $E_N = 0.25$ eV. To take account of the effects due to long and medium range surface sites correlation a large surface lattice is necessary, such as the 100×100 lattice used in simulating figure 9.10, since it is known the correlation affects the surface morphologies profoundly.

9.5.6   Dynamical RHEED from MBE growing surfaces

In principle diffuse diffraction patterns may be simulated by using the super unit cell method [59], performing many-beam dynamical RHEED calculations for a large number of surface configurations such as the one shown in figure 9.10 and then carrying out statistical average over these configurations. An estimation of the computation effort involved shows that this is in practice an impossible task for any realistic models of MBE growth. Taking a simple surface of an f.c.c. structure as an example, typically one needs to use more than 50 rods to achieve a convergent result for many-beam RHEED calculations and at least ten rods to gain an estimation of the values of diffracted beam intensities. Dynamical RHEED calculations using a super unit cell of a 100×100 lattice (see Figure 9.10) would involve at least a few thousand to a few million diffracted beams, and this is not only infeasible but also undesirable since most of the time the atomic details of reconstructed growing surfaces are unknown. In what follows we shall therefore concentrate on a simple approach [52, 50, 60] in which the detailed atomic structures of growing surfaces are not required and the computation may be performed using even a personal computer. Almost all features of RHEED intensity oscillations observed on the low index surface are reproduced using this simple approach. Within the validity of the approach, assuming that the range of correlation functions characterizing the surface disorder is relatively small, we will show that RHEED intensity depends only on the surface layer coverages, and that RHEED intensity oscillations are caused by the variations of the growing surface layer coverages. The detailed lateral structure of surface islands, although not affecting directly the RHEED intensity, influences the transfer of adatoms in between the growing layers and therefore indirectly the time evolution of RHEED intensity, that is the overall shape of the RHEED intensity oscillation curves.

We start our development by dividing the time-dependent scattering potential V into two parts, i.e. the time-averaged part $\overline{V}$ and the deviation of the real potential from the time-averaged potential δV. Since the time-averaged potential is time independent, scattering of the fast electrons by this potential does not involve any energy losses and is therefore called *elastic scattering*. Within our approach the elastic scattering is treated fully. The deviation potential δV, on the other hand, is in general time dependent. Scattering by this potential involves energy losses and is often called *inelastic scattering*. For disorder scattering since the energy losses involved are very small and since scattering by the aperiodic part of the potential δV gives rise to the appearance of diffracted beams in between Bragg spots, in what follows we shall refer to scattering by δV as *diffuse scattering*.

It has now been firmly established in transmission electron diffraction (TED) that to an excellent approximation the effect of the diffusely scattered electrons on the elastic wave field may be treated using a first order perturbation. In this approximation the effective scattering potential may be regarded as being complex and is called the *optical potential*. Assuming that the surface disorders are of short range nature, it has been shown (see for example [60]) that to a good approximation the optical potential occurring in the wave equation and consequently the elastic wave function are independent of the detailed distribution of the surface disorders parallel to the surface, i.e. surface step density. In other words as long as the surface islands are small *elastic diffracted electrons carry no information concerning the detailed atomic arrangement of adatoms within a surface layer. Only the layer coverage, or the distribution of atoms among the various layers, can be retrieved from the measurement of elastically diffracted beam intensities.* To be more explicit we may divide the optical potential into three parts

$$V(r) = V^{(1)}(r) + V^{(2)}(r) + V^{(3)}(r) \tag{9.60}$$

in which the first term on the right-hand side represents the averaged potential of the lattice, the second term results from the effects of thermal diffuse scattering on the elastic wave field and the third term denotes contributions from the diffusely scattered beams due to the surface disorders. All three terms depend only on the surface layer coverages θ_i. On the other hand if the growing surfaces are dominated by large islands, i.e. the correlation radius characterizing the surface disorder is relatively large, the reflected beam intensities are likely to show dependence on the detailed distribution of islands on the surface, i.e. surface step distribution and density, and in some limiting cases RHEED intensity oscillations might be dominated by contributions resulting from surface steps [61, 28].

In conjunction with the conclusion that only the surface layer coverages are the retrievable quantities in elastic RHEED experiments, we choose to use a so called *systematic reflection approximation* or one-rod RHEED approximation in which only surface layer coverage is involved. The basic idea behind this approximation is demonstrated in figure 9.11 for the AlAs($\bar{1}$00) surface. In figure 9.11(a) it is shown that when viewing the crystal lattice along the [001] high symmetry zone axis, the lattice periodicity parallel to the surface, i.e. along [010], is important and not negligible. On the other hand when viewing the same lattice a few degrees away from the main [001] zone axis figure 9.11(b) shows that the lattice periodicity parallel to the surface or along the [010] direction is no longer important, meaning that only the periodicity along the surface normal, i.e. [100] direction, or scattering by planes parallel to the surface are important. Neglecting crystal periodicity parallel to the surface and considering only scattering by planes parallel to the surface we then arrive at the *systematic diffraction approximation* of RHEED. The validity of this approximation is

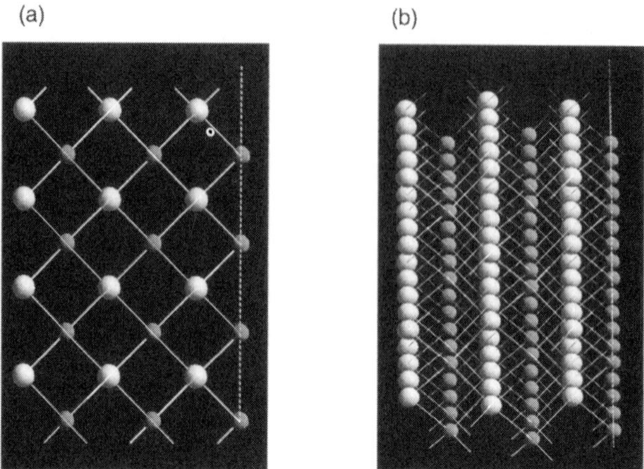

Figure 9.11 Atomic model of an AlAs lattice viewing along (*a*) the [001] zone axis and (*b*) a few degrees away from the zone axis. The surface is marked by the dotted line.

supported by many-beam RHEED calculations such as the one presented in figure 9.12, showing that while reflected beam intensities are very different between many-beam calculations for a [110] zone axis diffraction geometry and the systematic reflection approximation, this approximation works excellently for a diffraction geometry where the crystal is tilted away from the [110] zone axis by about 30 mrad.

Shown in figure 9.13 is a simulated RHEED intensity evolution curve for the homoepitaxial growth of GaAs and the specular reflected beam, using a distributed growth model (see below) with $A = 0.90$. The primary beam of energy 20 keV is incident on the growing surface at 50 mrad. Almost all features of measured intensity oscillations from a low index surface can be reproduced (for details see [50, 62, 63]). The distributed growth model is one of the birth–death models first proposed by Cohen and co-workers [56] for the homoepitaxial growth. Generalization to heteroepitaxial growth is straightforward and the models may be utilized to investigate such aspects of RHEED intensity oscillations as the measurement of growth rate and alloy composition, interface and recovery effects and control of quantum well widths [63].

Since on a low index surface there is no natural spatial origin, a mean field description is possible. By assuming that the net diffusion of the ith species from an upper layer to the layer immediately below is proportional to a product of the mobile ith (i.e. uncovered) species on the upper layer and the number of sites which are uncovered in the second layer below, the layer coverage $\theta_n^{(i)}$

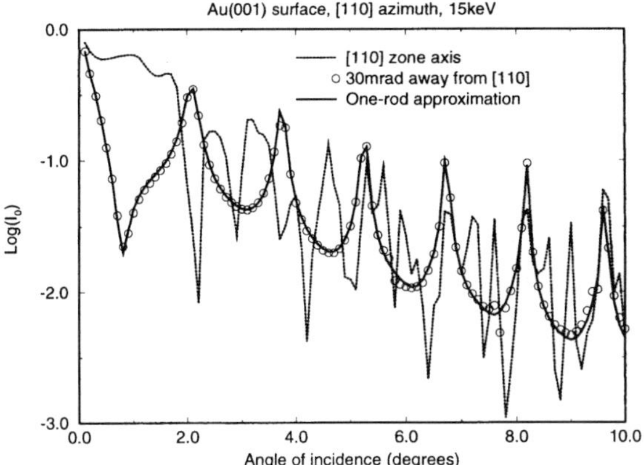

Figure 9.12 Specular reflected beam amplitude calculated using (- - - -) 21 rods and (——) one rod for the Au(001) surface.

of the ith species on the nth layer then satisfies a set of non-linear differential equations

$$\frac{d\theta_n^{(i)}}{dt} = d^{(i)}(t) \sum_j [\theta_{n-1}^{(j)} - \theta_n^{(j)}]$$

$$+ k^{(i)} \left[\theta_{n+1}^{(i)} / \sum_j \theta_{n+1}^{(j)} \right] \sum_j [\theta_{n+1}^{(j)} - \theta_{n+2}^{(j)}] \sum_j [\theta_{n-1}^{(j)} - \theta_n^{(j)}]$$

$$- k^{(i)} \left[\theta_n^{(i)} / \sum_j \theta_n^{(j)} \right] \sum_j [\theta_n^{(j)} - \theta_{n+1}^{(j)}] \sum_j [\theta_{n-2}^{(j)} - \theta_{n-1}^{(j)}]. \tag{9.61}$$

This set of equations gives the diffusive growth model of epitaxial growth, in which $d^{(i)}$ is the deposition rate for the ith species, and $k^{(i)}$ is a diffusion parameter. An alternative to the diffusive growth model is to treat the transfer of atoms between layers in accordance with the lateral structure of the layers [56]. In this model we assume that among all the ith species arriving on top of the nth layer per unit time, only a fraction $\alpha_n^{(i)}$ will transfer to the nth layer below and a fraction of $1 - \alpha_n^{(i)}$ will remain on top of the nth layer. The set of non-linear rate equations then takes the form

$$\frac{d\theta_n^{(i)}}{dt} = d^{(i)}(t) \sum_j [\theta_{n-1}^{(j)} - \theta_n^{(j)}]$$

$$+ \alpha_n^{(i)} d^{(i)}(t) \sum_j [\theta_n^{(j)} - \theta_{n+1}^{(j)}] - \alpha_{n-1}^{(i)} d^{(i)}(t) \sum_j [\theta_{n-1}^{(j)} - \theta_n^{(j)}] \tag{9.62}$$

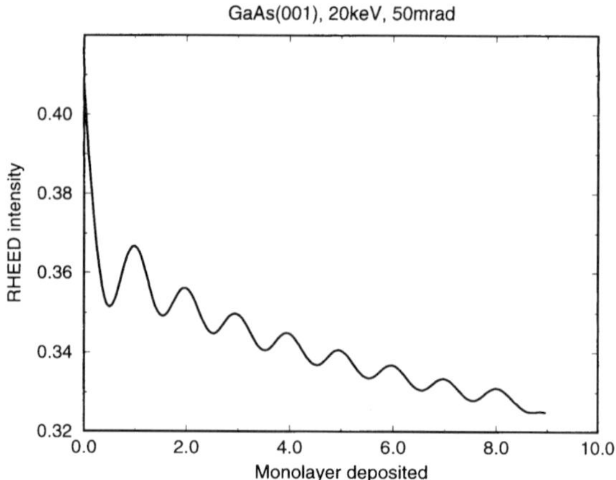

Figure 9.13 Simulated RHEED intensity oscillations during distributed homoepitaxial growth on a GaAs(001) substrate.

in which

$$\alpha_n^{(i)} = A^{(i)} \frac{p_n}{p_n + p_{n+1}} \tag{9.63}$$

and p_n is the total perimeter of the nth layer. The main advantage of this model is that there is considerable flexibility in defining the dependence of the perimeter on the coverage, and a simple approximation is to define

$$p_n \propto \sum_j \theta_n^{(j)} \left[1 - \sum_j \theta_n^{(j)} \right]^{1/2}. \tag{9.64}$$

In calculating figure 9.13, equations (9.63) and (9.64) were employed and the constant A was assumed to be 0.90. It should be pointed out that while for MBE growth involving only short range order disorders or small surface islands, such as those shown in figure 9.10, RHEED intensity depends only on the layer coverage rather then the detailed lateral distribution of surface disorders and therefore surface step density, the time evolution of RHEED intensity depends explicitly on the lateral structure of the growing layers via the transfer rate dependence on surface step distribution of adatoms in between different growing layers. RHEED intensity oscillations may therefore be used to retrieve parameters which govern the islands' evolution on the same layer.

9.6   SUMMARY

A dynamical theory of RHEED is developed within the framework of Bloch wave formulation. The original theory of Bethe is first introduced, and its advantages and disadvantages are discussed. The Bethe theory is then reformulated in terms of two-dimensional Bloch waves, and a numerical procedure is given for obtaining three-dimensional Bloch waves existing in a semi-infinite bulk crystal from two-dimensional Bloch waves. Approximate methods are also examined, in particular the one-rod or systematic reflection approximation and the Bethe potential method. Finally, dynamical RHEED theory is combined with Monte Carlo (MC) and birth–death models of molecular beam epitaxy (MBE) growth of thin films. Dynamical calculations for RHEED from MBE growing surfaces are performed and the origin of RHEED intensity oscillations and the use of RHEED for *in situ* monitoring and controlling of MBE growth are discussed.

We conclude that both elastic and diffusely scattered electron beam intensities oscillate during MBE growth. For short range surface disorders, e.g. small surface islands, the elastically scattered electron beam intensity depends only on the layer coverages of the growing surface layers and RHEED intensity oscillations result mainly from the periodic variation of the growing surface layer coverages. While under the short range surface disorder approximation the absolute RHEED intensity does not depend explicitly on the detailed lateral distribution of surface islands, the time evolution or the shape of the RHEED intensity oscillation curve may depend on the varying step configurations via its dependence on such parameters as the transfer rate of adatoms in between growing layers. During MBE growth, as the growing front becomes more and more diffused, i.e. as surface roughness increases, both RHEED intensity oscillation amplitude and the averaged intensity decrease. For the case of long range surface disorder or large islands on the surface RHEED intensity may depend directly on the lateral distribution of surface disorders and RHEED intensity oscillations may result mainly from the variation of surface step density during MBE growth.

ACKNOWLEDGMENTS

The author wishes to thank Dr S L Dudarev and Professor M J Whelan for collaboration and stimulating discussions over the years. This work is supported by the Chinese Academy of Sciences and National Natural Science Foundation of China.

REFERENCES

[1]   Davisson L and Germer L H 1927 The scattering of electrons by a single crystal of nickel *Nature* **119** 558–60
[2]   Pendry J B 1974 *Low Energy Electron Diffraction* (New York: Academic)
[3]   Duke C B (ed) 1994 *Surface Science—The First Thirty Years* (Amsterdam: North-Holland)

[4] Larsen P K and Dobson P J (ed) 1988 *RHEED and Reflection Electron Imaging of Surfaces* (New York: Plenum)

[5] Cowley J M 1992 Scattering factors for the diffraction of electrons by crystalline solids ed A J C Wilson *International Tables for Crystallography vol C* (Dordrecht: Kluwer) pp 223–44

[6] Bethe H 1928 Theorie der Beugung von Elektronen an Kristallen *Ann. Phys., Lpz.* **87** 55–129

[7] von Laue M 1931 The diffraction of an electron-wave at a single layer of atoms *Phys. Rev.* **37** 53–9

[8] Peng L-M and Cowley J M 1986 Dynamical diffraction calculations for RHEED and REM *Acta Crystallogr.* A **42** 545–52

[9] Peng L-M and Cowley J M 1988 A multislice approach to the RHEED and REM calculation *Surf. Sci.* **199** 609–22

[10] Ma Y 1991 A computational method for obtaining stationary solutions in RHEED and REM *Acta Crystallogr.* A **47** 137–9

[11] Anstis G R and Gan X S 1994 Calculation of RHEED intensities by a THEED algorithm *Surf. Sci.* **314** L919–24

[12] Gotsis H J and Maksym P A 1997 Investigation of multislice RHEED calculations *Surf. Sci.* **385** 15–23

[13] Strading R A and Klipstein P C (ed) 1990 *Growth and Characterization of Semiconductors* (Bristol: Hilger)

[14] Joyce B A 1985 Molecular beam epitaxy *Rep. Prog. Phys.* **48** 1637–97

[15] Hirsch P B, Howie A, Nicholson R N, Pashley D W and Whelan M J 1977 *Electron Microscopy of Thin Crystals* (Malabar: Krieger)

[16] Humphreys C J 1979 The scattering of fast electrons by crystals *Rep. Prog. Phys.* **42** 1825–87

[17] 1993 *The NAG Fortran Library Manual, Mark 16* Numerical Algorithms Group, Oxford

[18] Smith B T, Boyle J M, Dongarra J J, Garbow B S, Ikebe Y, Klema V C, and Moler C B 1976 *Matrix Eigensystem Routines—EISPACK guide* (Berlin: Springer)

[19] Colella R 1972 N-beam dynamical diffraction of high energy electrons at glancing incidence. General theory and computational methods *Acta Crystallogr.* A **28** 11–5

[20] Ashcroft N W and Mermin N D 1976 *Solid State Physics* (Philadephia, PA: Saunders)

[21] Peng L-M 1989 A note on the general Bloch wave theory and boundary problems in RHEED and REM *Surf. Sci.* **222** 296–312

[22] Peng L-M and Whelan M J 1990 A general matrix representation of the dynamical theory of electron diffraction—I. General theory *Proc. R. Soc.* A **431** 111–23

[23] Maksym P A and Beeby L 1981 A theory of RHEED *Surf. Sci.* **110** 423–38

[24] Ichimiya A 1983 Many-beam calculation of reflection high energy

electron diffraction (RHEED) intensities by the multi-slice method *Japan. J. Appl. Phys.* **22** 176–80

[25] Smith A and Lynch D F 1988 Results of multislice matrix calculations for convergent-beam RHEED patterns *Acta Crystallogr.* A **44** 780–8

[26] Zhao T C, Poon H C and Tong S Y 1988 Invariant-embedding R-matrix scheme for reflection high energy electron diffraction *Phys. Rev.* B **38** 1172–82

[27] Peng L-M, Dudarev S L and Whelan M J 1996 Dynamical RHEED calculations from the surface of a semi-infinite crystal *Acta Crystallogr.* A **52** 471–5

[28] Dudarev S L 1997 Incoherent scattering of electrons by a crystal surface *Micron* **28** 139–58

[29] Doyle P A and Turner P S 1968 Relativistic Hartree–Fock X-ray and electron scattering factors *Acta Crystallogr.* A **24** 390–7

[30] Peng L-M, Ren G, Dudarev S L and Whelan M J 1996 Robust parameterization of elastic and absorptive electron atomic scattering factors *Acta Crystallogr.* A **52** 257–76

[31] Peng L-M 1998 Parameterization of electron scattering factors of ions *Acta Crystallogr.* A **54** 481–5

[32] Tournarie M 1962 Recent developments of the matrical and semi-reciprocal formulation in the field of dynamical theory *J. Phys. Soc. Japan* **17** (Supplement B-II) 98–100

[33] Kambe K 1967 Theory of electron diffraction by crystals *Z. Naturf.* **22** 422–31

[34] Peng L-M, Ren G, Dudarev S L and Whelan M J 1996 Debye–Waller factors and absorptive scattering factors of elemental crystals *Acta Crystallogr.* A **52** 456–70

[35] Mott N F and Massey H S W 1965 *The Theory of Atomic Collisions* (Oxford: Clarendon)

[36] Rez D, Rez P and Grant I 1994 Dirac–Fock calculations of x-ray scattering factors and contributions to the mean inner potential for electron scattering *Acta Crystallogr.* A **50** 481–97

[37] Peng L-M, Dudarev S L and Whelan M J 1998 Electron scattering factors of ions and dynamical RHEED calculations from surfaces of ionic crystals *Phys. Rev.* B **57** 7259–65

[38] Peng L-M 1997 Anisotropic thermal vibrations and dynamical electron diffraction by crystals *Acta. Crystallogr.* A **53** 663–6

[39] Peng L-M, Dudarev S L and Whelan M J 1997 The ionicity of nickel monoxide: direct determination by reflection diffraction of high energy electrons from the (100) surface *Phys. Rev.* B **56** 15 314–9

[40] Tasker P W 1979 The surface properties of uranium dioxide *Surf. Sci.* **78** 315–24

[41] Herman M A and Sitter H 1988 *Molecular Beam Epitaxy—Fundamentals and Current Status* (Berlin: Springer)

[42]   Harris J J, Joyce B A and Dobson P J 1981 Oscillations in the surface structure of Sn-doped GaAs during growth of MBE *Surf. Sci.* **103** L90–6

[43]   Wood C E C 1981 RED intensity oscillations during MBE of GaAs *Surf. Sci.* **108** L441–3

[44]   Joyce B A, Dobson P J, Neave J H, Woodbridge K, Zhang J, Larsen P K and Bolger B 1986 RHEED studies of heterojunction and quantum well formation during MBE growth—from multiple scattering to band offsets *Surf. Sci.* **168** 423–38

[45]   Vvedensky D D and Clarke S 1990 Recovery kinetics during interrupted epitaxial growth *Surf. Sci.* **225** 373–89

[46]   Shitara T, Vvedensky D D, Wilby M R, Zhang J, Neave J H and Joyce B A 1992 Step-density variations and reflection high energy electron diffraction intensity oscillations during epitaxial growth on vicinal GaAs(001) *Phys. Rev.* B **46** 6815–24

[47]   Korte U and Maksym P A 1997 Role of the step density in reflection high energy electron diffraction: questioning the step density model *Phys. Rev. Lett.* **78** 2381–4

[48]   Lent C S and Cohen P I 1984 Diffraction from stepped surfaces—I. Reversible surfaces *Surf. Sci.* **139** 121–54

[49]   Zhang J, Neave J H, Dobson P J and Joyce B A 1987 Effects of diffraction conditions and processes on RHEED intensity oscillations during the MBE growth of GaAs *Appl. Phys.* A **42** 317–26

[50]   Peng L-M and Whelan M J 1991 Dynamical calculations for RHEED from MBE growing surfaces—I. Growth on a low-index surface *Proc. R. Soc.* A **432** 195–213

[51]   Mitura Z, Dudarev S L and Whelan M J 1998 The phase of RHEED oscillations *Phys. Rev.* B **57** 6309–12

[52]   Peng L-M and Whelan M J 1990 Dynamical RHEED from MBE growing surfaces *Surf. Sci. Lett.* **238** 446–52

[53]   Mitura Z, Mazurek P, Paprocki K, Mikolajczak P, Maksym P A and Beeby J L 1995 Structural characterization of thin films with the use of RHEED azimuthal plots *Inst. Phys. Conf. Ser.* vol 146 (Bristol: Institue of Physics) pp 541–4

[54]   Horio Y and Ichimiya A 1994 Origin of phase shift phenomena in RHEED intensity oscillation curves *Surf. Sci.* **55** 321–8

[55]   Madhukar A and Ghaisas V 1988 The nature of molecular beam epitaxial growth examined via computer simulations *CRC Crit. Rev. Solid State Mater. Sci.* **14** 1–130

[56]   Cohen P I, Petrich G S, Pukite P R, Whaley G J and Arrott A S 1989 Birth–death models of epitaxy—I. Diffraction oscillations from low index surfaces *Surf. Sci.* **216** 222–48

[57]   Kariotis R and Lagally M G 1989 Rate equation modelling of epitaxial growth *Surf. Sci.* **216** 557–78

[58] Weeks J D and Gilmer G H 1979 Dynamics of crystal growth *Adv. Chem. Phys.* **40** 157–228

[59] Kawamura T and Maksym P A 1985 RHEED from stepped surfaces and its relation to RHEED intensity oscillations observed during MBE *Surf. Sci.* **161** 12–24

[60] Dudarev S L, Peng L-M and Whelan M J 1992 A treatment of RHEED from a rough surface of a crystal by an optical potential method *Surf. Sci.* **279** 380–94

[61] Dudarev S L, Vvedensky D D and Whelan M J 1994 Statistical treatment of dynamical electron diffraction from growing surfaces *Phys. Rev.* B **50** 14 525–38

[62] Peng L-M and Whelan M J 1991 Dynamical calculations for RHEED from MBE growing surfaces—II. Growth interruption and surface recovery *Proc. R. Soc.* A **435** 257–67

[63] Peng L-M and Whelan M J 1991 Dynamical calculations for RHEED from MBE growing surfaces—III. Heteroepitaxial growth and interface formation *Proc. R. Soc.* A **435** 269–86

[64] Van Hove J M, Lent C S, Pukite P R and Cohen P I 1983 Damped oscillations in reflection high energy electron diffraction during GaAs MBE *J. Vac. Sci. Technol.* B **1** 741–6

INDEX